BIOLOGY OF TURBELLARIA: EXPERIMENTAL ADVANCES II

Papers by
Teiichi Betchaku, Gordon Marsh, Krystyna,
D. Ansevin, Christopher S. Lange et al.

MSS Information Corporation
655 Madison Avenue, New York, N.Y. 10021

Library of Congress Cataloging in Publication Data
Main entry under title:

Biology of Turbellaria.

    1. Turbellaria--Addresses, essays, lectures.
I. Shapira, Jacob, 1926-   [DNLM: 1. Turbellaria
--Cytology--Collected works. QX 350 B615 1973]
QL391.T9B5   595'.123'04108   72-13502
ISBN 0-8422-7085-X

TABLE OF CONTENTS

CREDITS AND ACKNOWLEDGMENTS

Ansevin, Krystyna D.; and Marilyn A. Wimberly, "Modification of Regeneration in *Dugesia tigrina* by Actinomycin D," *The Journal of Experimental Zoology*, 1969, 172:349-362.

Betchaku, Teiichi, "Isolation of Planarian Neoblasts and Their Behavior *in Vitro* with Some Aspects of the Mechanism of the Formation of Regeneration Blastema," *The Journal of Experimental Zoology*, 1967, 164:407-434.

Betchaku, Teiichi, "The Cellular Mechanism of the Formation of a Regeneration Blastema of Fresh-water Planaria, *Dugesia dorotocepha*. I. The Behavior of Cells in a Tiny Body Fragment Isolated *in Vitro*," *The Journal of Experimental Zoology*, 1970, 174:253-280.

Lange, C. S., "Observations on Some Tumours Found in Two Species of Planaria-*Dugesia etrusca* and *D. ilvana*," *Journal of Embryology and Experimental Morphology*, 1966, 15:125-30.

Lange, C. S., "A Possible Explanation in Cellular Terms of the Physiological Ageing of the Planarian," *Experimental Gerontology*, 1967, 3:219-30.

Lange, C. S., "Studies on the Cellular Basis of Radiation Lethality: I. The Pattern of Mortality in the Whole-body Irradiated Planarian (Tricladida, Paludicola)," *International Journal of Radiation Biology* 1968, 13:511-530.

Lange, C. S., "Studies on the Cellular Basis of Radiation Lethality: II. Survival-curve parameters for Standardized Planarian Populations," *International Journal of Radiation Biology*, 1968, 14:119-132.

Lange, C. S., "Studies on the Cellular Basis of Radiation Lethality: IV. Confirmation of the Validity of the Model and the Effects of Dose Fractionation," *International Journal of Radiation Biology*, 1968, 14:539-51.

Lange, C. S., "Studies on the Cellular Basis of Radiation Lethality: V. A Survival Curve for the Reproductive Integrity of the Planarian Neoblast and the Effect of Polyploidy on the Radiation Response," *International Journal of Radiation Biology*, 1968, 15:51-64.

Lange, C. S.; and C. W. Gilbert, "Studies on the Cellular Basis of Radiation Lethality: III. The Measurement of Stem-cell Repopulation Probability," *International Journal of Radiation Biology*, 1968, 14:373-88.

Marsh, Gordon, "The Effect of AC Field Frequency on the Regeneration Axis of *Dugesia tigrina*," Growth, 1969, 33:291-301.

PREFACE

Turbellaria are the most primitive living bilateral metazoans. Their close evolutionary relation to more advanced protostomes is evident in detailed similarities of their early embryonic processes, e.g. spiral cleavage, and mode of formation of mesoderm. Yet the simplest Turbellarians, the Acoels, display a number of special characteristics similar to certain groups of ciliate protozoans. Observations of the Acoel have led to the proposal that they occupy a phylogenetic "missing link" position. This fascinating group has, therefore, continued to be the subject of much experimental research during the last decade.

This collection of research reports deals with the molecular and cell biology of Turbellarians and ultrastructural studies on their specific cell types. Papers also discuss Turbellarian physiology and development as well as their regeneration and behavioral conditioning. Reports on radiation effects in Turbellaria complete the volume.

Regeneration

Isolation of Planarian Neoblasts and Their Behavior in Vitro with some Aspects of the Mechanism of the Formation of Regeneration Blastema

TEIICHI BETCHAKU

In spite of the numerous studies made since the turn of this century, the origins and properties of the cells which form the regeneration blastema of planaria are still uncertain; so also are the mechanisms underlying the histogenesis of the blastema. At present, most workers (Brøndsted, '55; Lindh, '57; Pedersen, '59) accept the hypothesis initiated by the work of Dubois ('49) and supported by the studies of the school of Et. Wolff (Dubois, Lender, Sengel and Wolff). From her x-ray studies, Dubois concluded that the blastema is formed entirely from free totipotent parenchyma cells or "neoblasts" which migrate to the wound. Lindh was in doubt about the distance which the neoblasts are capable of migrating, although Dubois ('49) and later Lender and Gabriel ('60) described their superior ability to migrate throughout the body. However, the complicated nature of planarian parenchyma and the difficulties of both fixing and staining tissues of this animal (Hyman, '51; Török and Röhlich, '59; Pedersen, '61a) made the previous studies inconclusive as far as the nature and the behavior of the blastema forming cells or the neoblasts are concerned. It must be emphasized also that the identity of the neoblasts themselves under a light microscope has not been fully established (Pedersen, '61a). Neither has the possibility of dedifferentiation of old cells, of their migration and of their redifferentiation, or a migration and multiplication of old cells, been disproved. The precise morphology and the cytochemistry of each type of planarian cells must be learned. The interaction between tissues or cells during the reconstitutional movement of regeneration has to be analyzed.

The isolation of certain cell types (neoblasts, fixed parenchyma cells and gastrodermal cells) and their characteristic morphology and their behavior in vitro will be dealt with here. The cytochemistry and ultrastructure of the isolated cells including nerve cells will be reported elsewhere.

Terminology

The terminology of planarian cells used in this paper follows the description of Pedersen ('59, '61a). The term, *neoblast*, is applied to the cell which has high RNA content in its scant cytoplasm without reference to its embryonic or morphogenetic relation to other cells. These cells are found abundantly in the regeneration blastema and throughout the parenchyma of the animal. They have been called previously "Stammzellen" (Keller, 1894), "Ersatzzellen" (Flexner, 1897), "formative cells" (Curtis, '02), "Regenerationszellen" (Bartsch, '23), "cellules libres du parenchyme" (Prenant, '22) and "neoblasts" (Buchanan, '33; Wolff and Dubois, '48) and are now most often called "neoblasts." Another type of parenchyma cells which, together with neoblasts, constitutes planarian parenchyma has large multiprocessed cytoplasm which is low in RNA content and high in PAS positive content (Pedersen, '61a). Following Pedersen, they are here designated as *fixed parenchyma cells*.

MATERIAL

Specimens of the triclad flatworm, *Dugesia dorotocephala,* were collected at a spring near Crawfordsville, Indiana. They were 1.6–2.0 mm wide and 12.1–16.8 mm long and were unfed under the laboratory conditions (17°–18°C, aerated) for a few days before the operation.

Culture media

Because of their superiority for microscopical observations and cytochemical studies, only liquid media were used in this work. Contrary to the results reported by other workers, if proper pH, temperature and salt concentration are provided, even isolated cells may survive up to eight days (tested with neoblasts and gastrodermal phagocytes) in plain saline solution without additional nutrients, provided there are other cells or fragments in the container. Since the purpose of this study is to identify isolated cells and to determine their behavior and early morphological changes *in vitro* within a few days after the culture is started, the effects of the saline solutions and nutrient media used were not critically tested beyond seven days. The improvement of the nutritive medium is left to future work.

Saline solution. Holtfreter's amphibian saline solution (Holtfreter, '31) was employed as a base. This saline solution was previously used in the work on the tonicity of the culture media for *Dugesia gonocephala* at Kyoto University (Betchaku, unpublished) and no deleterious effect was observed on culture tissue fragments. To determine the proper strength to be used, a rough test was first made by watching the reaction of the tissues near a wound, particularly the muscular contraction at the wound area, to saline solutions of a varied strength. That concentration of saline solution in which muscular contraction of the wound area did not occur and in which there was no immediate sign of shrinkage or swelling of the cells at the wound was chosen as most suitable. This was half strength Holtfreter's solution. Then, with neighboring concentrations, a survival test of isolated cells was made. The optimum dilution was found to be five-eighths concentration of the original Holtfreter's solution. Temperature and pH tests were not made critically, but throughout the experiments they were kept at 17°–18°C and at around pH 7.4. The pH range between 7.2 to 7.4 appears favorable for the cells. Unless specified, the saline solution in this paper indicates five-eighths concentration of the original Holtfreter's solution (table 1).

Nutrient media. Eagle's medium (modified as shown in table 2) was employed as the basal culture medium for this work. As natural ingredients, chicken embryo extract, chicken serum, planarian tissue extract, and calf serum were tested in preliminary experiments. Each of these natural ingredients seemed to have a slightly different effect on planarian cells. However, calf serum was used routinely because of its convenience and economy.

TABLE 1

The saline solution used in this work

NaCl	2,188 mg
KCl	31 mg
CaCl₂	63 mg
NaHCO₃	125 mg
H₂O	1,000 ml

TABLE 2

The basal medium used in this work

	Milligrams per 1000 ml
L-Arginine HCl	5.6
L-Cystine	2.0
L-Histidine	1.1
L-Isoleucine	8.7
L-Leucine	4.7
L-Lysine	6.1
L-Methionine	2.7
L-Phenylalanine	2.8
L-Threonine	4.0
L-Tryptophan	0.7
L-Tyrosine	6.0
L-Valine	3.9
L-Glutamine	48.9
Choline Cl	0.3
Nicotinamide	0.3
Ca-D-Pantothenate	0.3
Pyridoxal HCl	0.3
Riboflavin	0.03
Thiamine HCl	0.3
i-inositol	0.3
Biotin	0.3
Folic acid	0.3
Sodium pyruvate	150.0
Glucose	300.0
Calf serum	3 ml
NaCl	1,750.0
KCl	25.0
$CaCl_2$	50.0
$NaHCO_3$	100.0
Neomycin	100.0

Sometimes calf serum (3%) was added to the saline solution which then was used as a culture medium. It could support the cells up to 100 hours (mitoses of isolated neoblasts were observed during this period) to provide sufficient time for studying the early behavior of the cells. When concentrations of calf serum of 1–3% were used in the basal medium the neoblasts tended to coalesce (fig. 1) and the cell borders became obscure. In saline solution and in the basal medium with 0.3% calf serum, the neoblasts were free and the cell borders were distinct.

Isolation of the neoblasts

No enzymatic treatment was employed to dissociate neoblasts from the fragments in the routine procedure. Free cells may be obtained from tiny fragments of planarian body by simply shaking them in saline solution. Planarian neoblasts adhere more strongly to solid substrates, such as glass and plastic surfaces, and survive better in

non-nutritive saline solution than other types of cells. When varied types of cells are freed in saline solution, the neoblasts adhere firmly to the bottom of the culture dish as soon as they settle. Most other cells have much less tendency to make firm attachment and remain floating in the medium (fig. 7). The muscle cells are next to the neoblasts in their adhesiveness to solid substrates, but they degenerate soon in saline solution (within 30 hours under our conditions). By washing out the free floating cells, the neoblasts can be isolated on the bottom of the culture dish. Not many cells are freed, however, from freshly made fragments by shaking. The intercellular binding of the fresh tissue fragments is tight. A high yield of neoblasts can be obtained when the fragments are pretreated in saline solution for 24 to 48 hours.

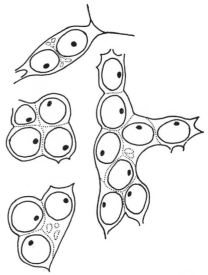

Fig. 1 When the amount of calf serum in the basal medium (table 1) was increased from 0.3% to 1%, neoblasts tended to coalesce. The cell borders are obscure. Both in the basal medium with 0.3% calf serum (table 2) and in the saline solution (table 1), neoblasts are free and the cell borders are distinct. Addition of calf serum (1% to 3%) alone to the saline solution does not cause the neoblasts to coalesce.

Neoblasts accumulate on the surface of a tiny fragment (fig. 2, 60) which has been kept in saline solution for 24 to 48 hours. Within 24 hours, a fragment usually becomes enveloped with migrated gastrodermis, but its intercellular binding is so fragile that by shaking the culture dish the gastrodermis breaks up into single spherical cells or small gastrodermal aggregates which float away into the medium. This exposes the neoblasts on the surface of the fragments. Since not only the gastrodermal cells but all intercellular connections become loosened, further shaking of these fragments releases many free cells, including neoblasts.

Step 4 and 5 of the procedure provide for (1) the removal of the cells injured by the mincing procedure or sudden contact with the saline solution; (2) the exposure of the accumulated neoblasts by breaking the envelope of gastrodermis, so that contact of the fragment with the bottom may result in the attachment of neoblasts to the cover glass on the bottom of a culture dish; (3) the removal of the cells released early (within 24 hours) which have a short survival period and are thus inadequate for the experiment. The short survival of cells released early may be due to insufficient time for their adaptation to the saline solution before being freed into it.

Fragments cultured in a nutrient medium immediately after being minced show no prominent accumulation of neoblasts on the surface. The intercellular bindings are not loosened except for the gastrodermal cells which slough into the medium as single spherical cells. Since they have less tendency to form a gastrodermal envelope over the fragment, more free gastrodermal cells are released in a nutrient medium. These problems will be discussed later.

Procedure

The procedure used in isolating the neoblasts from any area of the worms is as follows:

1. Clean the worms in autoclaved spring water with 0.02% Neomycin sulfate with several changes of fresh solution during 10 to 24 hours.

2. Mince the required area of the worms into tiny fragments (0.1 to 0.03 mm^3 or smaller) on a paraffin-wax block in one or two drops of the saline solution.

3. With a pipette, transfer the fragments into a culture dish filled with the saline soltuion. Two milliliters of saline solution in a 35 mm diameter sterile plastic petri dish (Falcon Plastics) is satisfactory for the amount of fragments

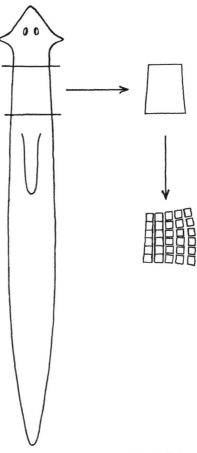

Fig. 2 The desired portion of the body is cut out in the saline solution and then minced into tiny fragments with a razor blade. They are used as the source of neoblasts for culture. In this work, only the prepharyngeal portion was used in order to standardize the source.

minced from the prepharyngeal regions of two worms of the size listed above.

4. After 24 hours, transfer the fragments with a pipette into the required number of petri dishes filled with the saline solution. Usually a square cover glass is placed on the bottom of the petri dish to facilitate either staining or subculturing the cells later. About 5 fragments in each dish containing 1.7 ml saline solution seems to be adequate.

5. After 24 hours, discard the saline solution and refill the dish with 1 ml of fresh saline solution or proper nutrient medium (see "Nutrient media"). Two fragments in each dish seems to be necessary for conditioning the fresh culture medium.

6. Examine each dish under an inverted microscope or a stereo dissecting microscope to determine the location of cell clusters, cell types and sizes. The majority of cells which have been attached to the bottom by this time are the neoblasts and many of the cell clusters or the monolayers are composed only of neoblasts. The neoblasts are now isolated and ready for further study. Since the attachment of the neoblasts to the bottom starts at step 4, observations on them can be made just after the fragments have been transferred into individual dishes. Note: the spring water used here was autoclaved, filtered and re-autoclaved after being bottled. The standard saline solution used was fiveeighths strength Holtfreter's amphibian saline solution with 0.01% Neomycin sulfate, pH 7.4. Sterile techniques should be carefully practiced at all times.

Fixing neoblasts on a cover glass for cytological study

1. *Cells growing on a cover glass.* A cover glass is placed in the culture dish before the cells are added. At desired intervals after the cells are attached to the cover glass, the medium is pipetted off and the fixative is added.

2. *Surface cells of a floating fragment.* The medium is removed until the bottom of the fragment touches the cover glass previously placed on the bottom of the petri dish. Then those cells which touch the cover glass are fixed by adding the fixative. The main mass of the fragment is removed by subsequent washings. An example of the results otbained with this technique is shown in figure 37.

If the medium is withdrawn completely before fixation in either of these procedures, the cells are greatly flattened and stretched by the surface tension of the medium as shown in figure 40. Fixed cells not distorted by artificial stretching are shown in figure 37–39. However, cytoplasmic structures may be examined better with flattened and stretched cells.

Identification of neoblasts and fixed parenchyma cells

The cytology of planarian cells either *in vivo* or *in vitro* has not been adequately studied. Murray ('27, '28, '31), Freisling and Reisinger ('58), Reisinger ('59), Seilen-Aspang ('57, '58, '60) and Ansevin and Buchsbaum ('61) left the identification of cells isolated *in vitro* rather unclear. This was because of the complexity of planarian parenchyma and the difficulty of both fixing and staining planarian cells (Pedersen, '59, '61a). While the primary focus of this paper is on neoblasts, none of the investigators of planarians has ever been able to make positive identification of neoblasts and nerve cells. There are several papers dealing with so-called "neoblasts," for example, those in which the number of neoblasts in the body has been counted (Brøndsted and Brøndsted, '61; Lender and Gabriel, '60; Stephane-Dubois, '61). However, the techniques employed in these studies did not enable the investigators to tell, *with certainty,* the difference between neoblasts and nerve cells except the most pronounced form of each cell types. The methyl green-pyronin staining used by these authors for the identification of neoblasts does not distinguish neoblasts from most nerve cells which also have a high RNA content in their sparse cytoplasm. They are similar in size and shape, except for the longer process (axon) of nerve cells which usually appears so extremely fine under the light microscope (down to 0.1–0.2 µ in diameter) that it is inadequate to use as one of the differentiating criteria of these cells in histological sections. In electron micrographs, the difference between these cell types are becoming clear (neoblasts: Pedersen, '59; nerve axon:

Török and Röhlich, '59; nerve cells: Betchaku, in preparation).

Under the circumstances, the identification of "neoblasts" in this paper was based on light microscopical criteria described by previous authors (Dubois, '49; Brøndsted, '55; Pedersen, '59), thus leaving some uncertainty about the identification of nerve cells. While the methyl green-pyronin method does not clarify definitely the difference between neoblasts and nerve cells, it certainly eliminates the confusion with all other cells. In this work, the observation of "neoblasts" started at a very early stage of isolation when "neoblasts" were spherical without processes. The brief observation of such round neoblasts without processes taken from an approximately six day old regeneration blastema of *Planaria vitta* has been reported by Pedersen ('59). When the neoblasts are isolated from a fragment in saline solution, they usually first take a spherical shape without processes (fig. 4). These cells were fixed with Zenker's fixative and stained with Lillie's Azure A and Eosin B method (Lillie, '54) at pH 5.0. Pedersen ('59) described this procedure as by far the best selective staining of the neoblasts; the cytoplasm of the neoblast is stained deep blue. Since the strong basophilia of the cytoplasm of the neoblast has been proven to be due to its high content of RNA (Clement, '44; Brøndsted, '55; Lindh, '57; Pedersen, '59), the methyl green and pyronin Y staining (Brachet, '42) was made on the fixed neoblasts accompanied with the extracting methods of RNA with RNAse (Brachet, '53) and with hydrochloric acid (Casselman, '59) (fig. 5–6). The cells which attached to the bottom surface of the culture dish satisfied the criteria of the neoblasts described by previous authors (Dubois, '49; Brøndsted, '55; Pedersen, '59).

By his electron microscopical and histochemical observations, Pedersen ('61a) reported that the fixed parenchyma cells of fresh-water planarians, *Planaria vitta* and *Dugesia tigrina,* are large and have many attenuated cytoplasmic processes with a high content of PAS positive material. To identify the fixed parenchyma cells other than by their morphology, the PAS test (Hotchkiss, '48) was employed.

In the sections (Zenker's fixation, paraffin) of intact worms, *Dugesia dorotocephala,* an intense PAS reaction of these cells was obtained. In the same preparations, the cytoplasm of both neoblasts and gastrodermal cells did not show a detectable reaction, although gastrodermal phagocytes often contained PAS positive inclusions in their cytoplasmic vacuoles. A PAS test on the isolated cells *in toto* without embedding process showed the same but much clearer pictures of the reaction.

In general, to avoid misidentification of each cell type *in vitro,* the observation of the cells *in vivo* was made on a tiny fragment pressed slightly under a cover glass into a monolayer after the trypsin treatment as employed by Pedersen ('61a) for *in vivo* observation. Thus it was possible to study the morphology of planarian cells both *in vivo* and *in vitro* on the same preparation because some isolated cells were always found around the fragment. Planarian cells isolated in the saline solution retained their characteristics.

Observation of cells in vitro

Most observations of living cells were made through an inverted microscope (Unitron BMIC microscope body with Zeiss optical system). The short term observation of high magnification was done with a regular phase contrast microscope (Zeiss photomicroscope) by placing the culture on a slide. A time-lapse 16 mm cinematographic apparatus (Electro-Mechanical Development Co. unit with a Cinekodak Special) was used for recording the long term movement of the culture and a 35 mm still camera was used for recording the details of the culture. Cover glass (Gold Seal, Clay-Adams Co.) was used mostly as the substrate for the isolated cells. It was cleaned with a detergent ("7X", Linbro Chemicals, Inc.). On a few occasions the bottom of a plastic culture dish (Sterile pack, Falcon Plastics of B-D Laboratories, Inc.) was used as the substrate. A glass surface seemed to provide slightly better attachment of cells in general.

Neoblasts. When neoblasts are isolated *in vitro,* they first assume a round shape (fig. 4). The measurements of Zenker-fixed spherical specimens are as follows:

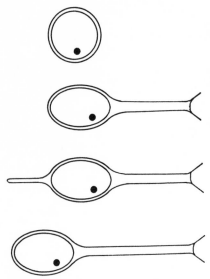

Fig. 3 When neoblasts are isolated *in vitro*, they first assume a round shape with sparse cytoplasm. Soon the cell develops one process (primary) and then another process (secondary) at the opposite pole. The third form is regarded as the basic form of neoblasts. Later, the secondary process degenerates leaving only the primray process which then becomes a long stable process, or more processes may be developed.

nuclei are round with a diameter range between 8.1 to 8.8 µ; cell bodies are also round with a diameter range between 8.8 to 10.8 µ. The cytoplasm is thus extremely thin, averaging 0.6 to 0.7 µ with a range of 2 µ to unmeasurable thickness. The nucleus usually contains one nucleolus of about 0.8 µ in diameter. Soon the cell develops a process (primary process) of around 8 to 11 µ in length which swings around. Being different from other types of cells, the movement of the neoblast depends strictly on this process which elongates, swings and contracts. Usually two tiny projections are seen at the tip of the process (fig. 3). When the tip of the process reaches an object and makes firm attachment with it, the process starts to contract and the cell body moves toward the tip of the conracting process. It soon develops another process (secondary process) at the opposite pole of the cell.

This bipolar shape of a neoblast with a large process at one pole and a small process at the opposite pole (fig. 8) seems to represent its basic form *in vitro*. Every freshly isolated neoblast of spherical form observed so far assumed this form within a couple of hours after attachment to the bottom of a culture dish. This form also seems to be the basic form of the neoblasts *in situ*. Histological sections of intact parenchyma (fig. 36) show almost identical form of neoblasts *in situ* which were regarded as the resting neoblasts by Pedersen ('59).

If the cell has not been participating in locomotive activity within 48 hours after isolation, it usually develops more processes (multipolar form), or the secondary process degenerates leaving only the primary process which then becomes a long stable process (unipolar form). These processes of secondary nature seem to lack the contractile strength equal to the primary process although they are motile.

The only observed method of locomotion of neoblasts *in vitro* was the contraction on the firmly attached processes *between two or more neoblasts* (figs. 28–29, 39, 41–42). The first contact of their processes was due to their randomly swinging primary processes. By physical contact only the neoblasts seem to recognize their own kind and strong mutual affinity was demonstrated between them by means of the firm attachment of the tips of their primary processes and their contraction. While the body surface of neoblasts is adhesive to the substrate (glass or plastic) more than that of any other types of cells in the culture, the first firm contact between neoblasts is made with the tips of their processes rather than by the cell bodies. The tip of the primary process of a neoblast did not make any firm attachment either to any other type of cells or to any debris of tissues of its own upon contact. Thus, if no other neoblast was within reach, no locomotion was observed: neoblasts do not move over or by other types of cells in "stepping-stone" fashion.

In this way, the locomotion of neoblasts always ends in their aggregation. An essential factor for aggregation *in vitro* is that the neoblasts must be within reach of each other's processes or must be brought within reach by other forces. When isolated neoblasts are close enough, and even when mixed among other types of cells, their aggregation takes place quite smoothly because of their low adhesiveness to other types of cells (fig. 7). When several neoblasts aggregate, some of them rotate their whole bodies and their primary processes extend outwards. The secondary processes seem to be used for binding or keeping the cell bodies in position. The secondary processes extending outwards degenerate (figs. 19–20).

In some instances of isolated neoblasts of either unipolar form or bipolar form, the cell body alone rotated by 180° while the process or processes remained stationary with the result that each process was then attached to the opposite pole of the cell body. During the rotation of the cell body, the process or the processes appeared feeble and motionless (figs. 15, 18). Each process regained its steady appearance and motility when the rotation was completed. In a bipolar neoblast, the rotation caused the characteristic transformation of the primary process to the secondary process and *vice versa*. The secondary process which had been thin and short then became thick and long and started to swing (figs. 11–16). In the case of a unipolar neoblast shown in figures 11–24, the appearance of the process did not change after the rotation of the cell body, but the motility of the process was considerably reduced.

In appearance and behavior the processes of neoblasts resemble the ectoplasmic filopodia observed in primary mesenchymal cells of sea-urchin embryos (Gustafson and Wolpert, '63). However, some of their unique features should be noted. These processes seem to be part of the basic structure of the neoblasts for they develop and retract in a certain regulated manner. When either cell body or process rotates against the other, the rotation ends at 180° and nowhere in between. This indicates that the bipolarity shown in their initial form has a structural foundation. Although these processes are present in the neoblasts of both intact (fig. 36) and regenerating planarians, ultrastructural studies of neoblasts (Pedersen, '59; Le Moigne et al., '65) disclosed no structural

difference between the perikaryon and the processes. Light microscopical observations of the fixed and stained neoblasts both *in vitro* and *in situ* showed that the basic cytoplasmic structure are homogeneous from the perikaryon through the process. The perikaryon and the process are equally granular and strongly basophilic (figs. 8, 36–40).

A step 6 of the isolation procedure, clusters ranging from a few cells to several hundred cells are to be found in various places on the bottom of a culture dish (fig. 25). Most of the clusters are freshly formed monolayers of the round neoblasts without processes. However, since the attachment of the neoblasts to the bottom starts as soon as the fragments are introduced into the culture dishes at step 4, in some clusters the primary processes of the cells may have developed and their swinging around may be seen; in some clusters the aggregation of the cells may already be under way.

Within 48 hours after a monolayer of the round neoblasts without processes has formed, either in plain saline solution or in nutrient medium, the cells pile up to form an aggregate (figs. 28–33). Because of their unique mechanism of locomotion which depends on the contraction of their mutually attached processes, the aggregations occur in various places of the sheet and later these small aggregates unite to make a larger aggregate. Whether two aggregates will fuse depends on the distance between the two aggregates as described earlier (fig. 27). When the aggregation is completed it is a spherical ball and a slight vibration or movement of the medium allows the aggregate to detach from the substrate and float in the medium (figs. 33–35). The histological study of this spherical aggregate will be considered in subsequent studies.

Lipid droplets were common in cultured neoblasts (figs. 11–26, 34, 41–49). They were stained heavily by the Oil Red O method (Lillie, '44) and easily extracted by Baker's Pyridine Extraction Test (Baker, '46). In cinematographic study, the secretion occurs in the perikaryon and the droplets tend to move along the process to the tip, with some backward and forward movement, and then suddenly to come

17

back to the perikaryon. The droplets fuse and redisperse frequently and continuously, indicating active protoplasmic movement in the process. Generally, an accumulation of lipid was observed at the site of the secretion (figs. 17–25). Each neoblast had only one place of secretion in its cytoplasm. In the bipolar and the unipolar neoblasts, the secretory site was usually at the proximal portion of the primary process. However, because of the rotation of the cell body as described earlier, the secretory site of an isolated unipolar neoblast was sometimes located in the perikaryon opposite to the process (figs. 10–12, 17–24). If the end of the process was inserted into the aggregate, the lipid droplets which had moved down the process sometimes did not return to the perikaryon. In the case of isolated neoblasts, the accumulation of lipid at the distal end of the process was observed (figs. 24–26). The fusion and redispersion of the droplets were repeated in the swollen distal end of the process. Before the cell started to degenerate, however, these droplets returned to the perikaryon and the process once again regained its normal appearance.

Fixed parenchyma cells. The described procedure for the isolation of neoblasts has been primarily designed to collect neoblasts selectively from any region of the body. The technique utilizes the differential susceptibility of planarian cells to culture conditions. The fixed parenchyma cells and other specialized cells such as various gland cells, muscle cells and flame cells are highly sensitive to the environmental factors. During cultivation in the saline solution, the fixed parenchyma cells gradually disappear from the exposed area of the parenchyma of the cultured fragments while the neoblasts accumulate in the same area. Histological sections of the fragments cultured in saline solution revealed the necrotic figures of the fixed parenchyma cells at the exposed surface of the parenchyma of the fragments (see: *Histological observation of the fragment*). However, by lightly pressing a cultured fragment on the bottom of the culture dish or on any desirable substrate, the fixed parenchyma cells from the deeper area may be attached to the substrate along with other types of cells. Because of their low survival rate in saline solution, more healthy fixed parenchyma cells are obtained from the fragments treated in a nutrient medium.

As shown in figure 50, the nuclei of the fixed parenchyma cells are smaller than those of the neoblasts. Their ample cytoplasm is so flexible that the form of the fixed parenchyma cells varies under different environmental conditions. They can be spherical while floating in a medium. They may spread a large thin sheet of hyaline ectoderm on a substrate when isolated alone, or produce numerous fine processes which extend through the gaps between cells when mixed with other cells. The cytoplasm has PAS positive content and many lipid droplets as reported by Pedersen ('61a).

An interesting property of the fixed parenchyma cells is their possible affinity with neoblasts. They may have similar affinity with other types of cells, but the fragility of these cells made direct observations of their interaction with other types of cells *in vitro* rather difficult. On a few fortunate occasions, their interaction with neoblasts *in vitro* was observed and photographed.

In the case of figures 41–49, the neoblasts and a fixed parenchyma cell were taken from a two day old fragment cultured in saline solution. They were then cultured in saline solution containing 3% calf serum. The fixed parenchyma spread a few very thin but wide sheet-like filopodia on the substrate (cover glass) and migrated by constantly expanding and retracting them. It is important to note that the fixed parenchyma cell actually pushed its way through the aggregating neoblasts from one side to the other. This indicates that both the locomotive and the structural strength of the fixed parenchyma cell exceed the contracting strength of the processes of the neoblasts or the adhesive strength between the neoblasts. The thin filopodia of the fixed parenchyma cell penetrated the interstices between the aggregating neoblasts and prevented them from making a tight aggregate. One of the neoblasts (marked by X in figs. 44–49) was separated from the aggregating neoblasts by the fixed parenchyma cell. The tips of neoblast processes did not show any adhesiveness to the fixed parenchyma cell.

These phenomena indicate what may actually be happening in the intact parenchyma. It should be the fixed parenchyma cells that prevent the neoblasts from aggregating in the parenchyma of an intact animal.

Gastrodermal cells. The gastrodermal cells which compose a single-layered epithelium of the intestine occupy a large portion of the body space. They are columnar and have many vacuoles. Gastrodermal cells isolated by treatment with 0.3% crude trypsin exhibit active amoeboid movement (figs. 51–57). The distinct separation of the cytoplasm into ectoplasm and endoplasm was observed in these moving gastrodermal phagocytes as reported by Pedersen ('61b). Their pseudopodia are always short, wide and rounded, and the cells change their shapes constantly. The rounded pseudopodia never seemed to adhere to the substrate firmly. The locomotion of isolated gastrodermal cells on the substrate (glass or plastic) consisted of a circling movement with a small movement of the cell body at each rotation. Some of the isolated gastrodermal cells, however, take on a spherical form (30–35 μ in diameter) and float in the medium. *In vitro,* no *fundamental* difference between so-called "granular clubs" and "phagocytic cells" (Hyman, '51) was observed; these forms may represent different stages of the same cell.

Multinucleated gastrodermal cells were often observed (fig. 58). A gastrodermal cell sometimes contains a small cell in one of its vacuoles (fig. 58). The nucleus of the trapped cells is usually smaller than that of the host cell and is surrounded by a very small amount of cytoplasm which separates into ectoplasm and endoplasm. The hyaline ectoplasm demonstrates an active waving motion around its perikaryon. The significance of this phenomenon is not known.

Histological observation of fragments cultured in saline solution

Figure 59 shows a middle section of a fragment when it was minced from the body in saline solution. While neoblasts can be collected from any part of the body,

all fragments used in this work were made from the prepharyngeal region to standardized the source (fig. 2). Each fragment contains almost all kinds of cells and tissues.

In fragments cultured in saline solution, two major cytological processes take place. One is the degeneration of the cells which are susceptible to this culture condition. The other is the aggregation of like-tissues or cells which survive the condition. The cells which are less susceptible are the gastrodermal cells, the epithelial cells and the neoblasts. Degenerating cells include the fixed parenchyma cells and the specialized types such as muscle cells, subepidermal pigment cells, varied types of gland cells and possibly protonephridial cells with their flame cells. The nerve cells degenerate at the very surface of the fragment but those inside the nervous plexus seem to be well protected and survive as long as the neoblasts do. Mitotic figures are, however, only observed among the neoblasts.

Figure 60 shows a middle section of a fragment which has been kept in saline solution for twenty-two and one-half hours. This is about the time when step 4 of the isolation procedure takes place. The gastrodermis has united into a continuous mass and covers the surface of the fragment. The cytoplasm of the gastrodermal cells become so extremely vacuolated that the cell borders are difficult to detect. This morphological change is quite different from the constancy in form of the isolated gastrodermal cells. The isolated gastrodermal cells do not show any significant change of their original appearance after twenty-two and one-half hours in saline solution. There is an increase in density of neoblasts toward the exterior of the interior mass of the fragment, and the considerable accumulation of neoblasts is observed around the surface of the interior mass. Degeneration of cells other than neoblasts in the parenchyma are apparent in the surface and subsurface area. Mitotic figures of neoblasts are observed.

The difference of the behavior between the cells in a fragment become more apparent when sections of a fifty and one-half hour old fragment cultured in saline solution are examined. The segregation

19

of unlike-tissues has almost been completed in the fragment of this group. In the fragment shown in figure 61, the gastrodermal tissue is completely separated. The main mass of this fragment shows an increase in the density of neoblasts among which mitotic figures are still observed along with the necrotic figures of the remaining other kinds of cells, including the fixed parenchyma cells. A middle section of another fifty and one-half hour old fragment (fig. 62) shows the separation of tissues. The original parenchyma, now composed almost entirely of neoblasts with extremely few fixed parenchyma cells, occupies one quarter of the fragment. The gastrodermis and the nervous tissue occupy two other separate quarters. The epithelial cells become clumped and occupy the remaining quarter. Tissues or cells other than those mentioned above are cytolysing. No dedifferentiating figures of cells are observed.

DISCUSSION

In the previous studies of the formation of the regeneration blastema of fresh-water planarians (Dubois, '49; Lender and Gabriel, '60; Lender, '62; Wolff, '62) the extensive directed migration of neoblasts toward the wound was discussed and emphasized. In the present work, however, both gastrodermal cells and fixed parenchyma cells were found to be actively migratory although each kind of cell showed individual characteristics both *in vitro* and *in vivo*. The neoblasts were also found as active mobile cells, but their unique mechanism of locomotion makes it improper to regard them as migratory cells capable of making extensive free migration either *in vitro* or *in vivo*.

Because of their extremely strong mutual affinity and almost non-existent affinity with other types of cells, the locomotion of neoblasts is achieved only by the mutual traction between the attached processes of nearby neoblasts, which will end in their local aggregation. In intact parenchyma, neoblasts are populous among fixed parenchyma cells (fig. 36). For a neoblast to make free migration at all, two conditions would be required. First, fixed parenchyma cells should provide sufficient adhesiveness to the tip of the primary process, so that the neoblast could have "stepping-stones" for its free migration. Second, a collison with another neoblast which would end in their aggregation should be avoided. The observations of the interaction between neoblasts and fixed parenchyma cells *in vitro* and *in vivo* (cultured fragments) were not in harmony with the first assumption. Even if a neoblast could migrate by using fixed parenchyma cells as "stepping-stones," its free path should be negligibly short because of the high population of neoblasts in the parenchyma.

Instead, the observations indicate that it is the fixed parenchyma cells that prevent the movement or the aggregation of neoblasts in the intact parenchyma. When the parenchyma is exposed to a hypotonic environment upon injury, the constituents of the parenchyma other than neoblasts would start to cytolyse thus freeing neoblasts. In the present work, the destruction of the fixed parenchyma cells immediately started a local aggregation of neoblasts. The treatment which avoids the destruction of fixed parenchyma cells prevented the aggregation of neoblasts.

The important characteristics of planarian cells (*Dugesia dorotocephala*) are (1) the great differential cellular susceptibility to environmental factors, especially to hypotonicity, (2) the great differential cellular affinity and (3) the great differential cellular mobility. These great cellular characteristics are probably three of the major factors responsible for the remarkable regenerative power of the animal because they provide an effective means for rejoining the fragmented tissues after injury.

These three factors are involved in the initial accumulation of neoblasts in a tiny fragment cultured in saline solution. The aggregation of gastrodermal cells, achieved by their active and rapid migration to the surface of the fragment, allows the divided parenchymal tissues to unite. Simultaneously, the cytolyses of the cells which fill the space between neoblasts in the parenchyma evoke the aggregation of neoblasts by freeing them and by bringing them closer to each other as a result.

The formation of an ordinary regeneration blastema against the fresh water environment and the accumulation of neoblasts on the surface of a fragment cultured in saline solution may not involve the same mechanism since in the former case wound closure occurs by the migration and the elongation of epithelial cells. In the latter case, the gastrodermal cells cover the surface of the fragment instead of the epithelial cells. However, the author's unpublished work on normal regeneration process of *Dugesia gonocephala* and the similar work recently made with *Dugesia dorotocephala* (Betchaku, in preparation) indicate that there is no basic difference in the mechanism of the *initial* accumulation of neoblasts at the wound in normal regeneration and at the surface of a cultured fragment. At the same temperature as for the *in vitro* experiments (17°–18°C), the covering of the wound by the epithelial cells takes place in one to two days in regular regeneration. Until that time, the wound area is sealed with the gastrodermal cells. Both the migration of gastrodermal cells either to the exterior (to the full extent in a cultured fragment and less in normal regeneration) or to the interior (in normal regeneration only) and the cytolyses of cells (in both cases) occur in this period. The muscle contraction around the wound area may contribute much to this movement of gastrodermis and the condensation of the parenchyma in which the cells susceptible to hypotonicity degenerate. During the reconstitutional movement of tissues, especially of the gastrodermis which may lead the movement of the parenchyma, it is also possible that the neoblasts in the nearby parenchyma make contact with the neoblasts of this accumulation and are pulled into it by their strong mutual affinity. However, mitoses may be the prime factor in increasing the population of neoblasts in a regeneration blastema as well as in a cultured fragment.

In a cultured fragment in saline solution, the accumulation of neoblasts or, in more appropriate words, the aggregations of like-tissues were completed by approximately 2 to 3 days. There was no sign of dedifferentiation or differentiation of cells in the fragment. Some minor morphological changes, mostly vacuolization, should be attributed to the process of the cellular deterioration in a non-nutritive medium. The rate of mitoses of neoblasts, their morphological appearance, their RNA content and their behavior pattern, if isolated *in vitro*, remained unchanged after four days of culture. This indicates that no differentiation of neoblasts had taken place in the fragment. The problems of the cellular interactions and cytodifferentiation in the regeneration blastema in connection with the totipotency of neoblasts will be discussed in a separate paper.

Whether or not the fixed parenchyma cells form a syncytium in the parenchyma of fresh-water planarians was well discussed with reference to the historical arguments in a paper on the nature of planarian connective tissue by Pedersen ('61a). He concluded from electron microscopic observations that the fixed parenchyma cells of *Planaria vitta* and *Dugesia tigrina* do not form a syncytium. Skaer ('61) in a similar study also reported the cellular nature of the parenchyma of *Polycelis nigra*. The present work confirms the cellular construction of the parenchyma. The fixed parenchyma cells are always isolated from the parenchyma as individual highly mobile cells. When a live fragment was observed by being pressed into a monalyer as described earlier, multinucleated fixed parenchyma cells were never observed. They were all *single nucleated* cells.

Whether the parenchyma is cellular or syncytial, only the structure of the parenchyma with *stationary* fixed parenchyma cells has been discussed by researchers. The fixed parenchyma cells demonstrated their vigorous migratory activity *in vitro* immediately after being taken from a fragment of the body. Their highly flexible processes and bodies have been observed to penetrate between the aggregating neoblasts and yet are strong enough to push them apart as shown in figures 41–49. Pedersen ('61a) suggested the possible mobility of the fixed parenchyma cells *in situ* by his electron microscopic observation that no desmosomes were found at the surfaces of these cells. Do the fixed parenchyma cells *in situ* retain such mobility as shown *in vitro*?

The neoblasts, another constituent of the parenchyma, showed their strong mutual affinity *in vitro* as they aggregated whenever their primary processes touched each other. The bipolar form of neoblasts *in situ* seen in the histological sections of the intact parenchyma was learned as their active mobile form by the observation of the isolated neoblasts *in vitro*. And yet, the neoblasts in the intact parenchyma do not aggregate. If the orientation of these polarized neoblasts *in situ* are taken as an indication of the direction of their movement, they are random as described by Lindh ('57) who predicted the mechanism of the locomotion of neoblasts from the ordinary histological sections as "by emitting fine protoplasmic threads, which after fastening are contracted." Do these indicate the general parenchymal movement of random nature or the mobility of the fixed parenchyma cells in the intact parenchyma?

There is some evidence of the existence of "parenchymal movement." First, Johnson and his associates ('62) observed that the subepidermal pigment of *Dugesia dorotocephala* moved to the gastrodermis and then was given off through the pharynx when the whole worm was treated in a medium containing an antifungal compound (Fungichromin) or certain unsaturated fatty acids. Although there was no report in the literature whether the pigment is intracellular or extracellular, the author's own observation of the pigment cells *in vitro* confirmed that the pigment is intracellular. Very likely the antifungal compound killed the pigment cells in Johnson's experiment. If such an antifungal compound or certain fatty acids had killed the pigment cells, what then caused the movement of the pigment to the gastrodermis? Second, there is a large population of neoblasts in the subepidermal area. However, in the parenchyma of the middle portion of the body, the density of neoblasts is not uniform. In some areas they are more numerous than in other areas. These two phenomena may be easily explained if a general movement of the fixed parenchyma cells in the parenchyma is assumed. If the function of the fixed parenchyma cells is the storage and the transport of the metabolites or wastes as suggested by Pedersen ('61a), the movement of fixed parenchyma cells in the paren-

chyma would be essential for their efficient function.

This "parenchymal movement" or the mobility of fixed parenchyma cells is a good explanation of the mechanism of the formation of the regeneration blastema where the local cellular material was not available from the damage caused by x-ray irradiation (Dubois, '49). Dubois misinterpreted the moving direction of a neoblasts by 180° in her diagramed illustration which was later quoted in a review paper of Wolff ('61) without criticism. Contrary to her claim, this mistake ironically proves that the accumulation of neoblasts at the site of regeneration in her experiment was not due to the migration of neoblasts through parenchyma, but that the neoblasts should have been carried to the site of regeneration by the fixed parenchyma cells, possibily with the association of the gastrodermal cells which are apt to migrate out through the open wound. The extended processes of neoblasts indicates that the strength of such carriage exceeds the contracting strength

of the processes of neoblasts or the adhesive strength between neoblasts. Indeed, the polarized shape of neoblasts which corresponds to their direction of movement enables the investigators to study direction of the relative movement of neoblasts and their surroundings on ordinary histological sections of regenerating planarians.

Flickinger ('64) tested the mobility of cells in the regenerating worms by grafting a piece of the isotopically labeled worm before the unlabeled host worm was decapitated. The immobility of the graft cells thus found was criticized by Best and his co-workers ('65) as the result of possible radiation injury due to high level of C^{14} employed. The regenerative capacity of the labeled worm should have been tested as the control.

ACKNOWLDEGMENT

The author is indebted to Professor M. Ichikawa, University of Kyoto, Japan and Professor H. V. Brøndsted, University of Copenhagen, Denmark, in whose laboratories the preliminary experiment of culturing technique of planarian tissues and the study of planarian cytology were made. He is also very grateful to Professor B. V. Hall, University of Illinois, Urbana, Illinois for his kind contesy of allowing him to use the facilities of his laboratory. He should like to thank the undergraduate students for their assistance on minor but indispensable aspects of the work for the project. The main part of the work was done in the laboratory of Professor W. H. Johnson with grants from the National Science Foundation, nos. GB303 and GB2888.

LITERATURE CITED

Ansevin, D., and R. Buchsbaum 1961 Observations on planarian cells cultivated in solid and liquid media. J. Exp. Zool., 146: 153–162.

Bartsch, O. 1923 Die Histogenese der Planarienregenerate. Arch Entw'mech., 99: 187–221.

Best, J. B., R. Rosenvold, J. Souders and C. Wade 1965 Studies on the incorporation of isotopically labeled nucleotides and amino acids in planaria. J. Exp. Zool., 159: 397–403.

Brachet, J. 1942 La localisation des acides pentosenucleiques dans les tissues animaux et les oeufs d'Amphibiens en voie de developpement. Arch. Biol., Paris, 53: 207–257.

—— 1953 The use of basic dyes and ribonuclease for the cytochemical detection of ribonucleic acid. Quart. J. Micros. Sci., 94: 1–10.

Brøndsted, H. V. 1955 Planarian regeneration. Biol. Rev., 30: 65–126.

Brøndsted, A., and H. V. Brøndsted 1961 Number of neoblasts in the intact body of Euplanaria torva and Dendrocoelum lacteum. J. Embr. Exp. Morph., 9: 167–172.

Buchanan, J. W. 1933 Regeneration in Phagocata gracilis (Leidy). Phys. Zool., 6: 185–204.

Casselman, W. G. B. 1959 Histochemical Technique. New York: John Wiley and Sons Inc., 1959.

Curtis, W. C. 1902 The life history, the normal fission and the reproductive organs of Planaria maculata. Proc. Boston Soc. Nat. Hist., 30: 515–559.

Dubois, Fr. 1949 Contribution a l'étude de la mitration des cellules de régénération chez les Planaries dulcicoles. Bull. Biol. Fr. Belg., 83: 213–283.

Flexner, S. 1897 The regeneration of the nervous system of Planaria torva and the anatomy of the nervous system of double headed form. J. Morph., 14: 337–346.

Flickinger, R. A. 1964 Isotopic evidence for local origin of blastema cells in regenerating planaria. Exp. Cell Res., 34: 403–406.

Freisling, M., and E. Reisinger 1958 Zur Genese and Physiologie von Restitutions Körpern aus Planarian-Gewebesbrei. Roux' Arch., 150: 581–606.

Gustafson, T., and L. Wolpert 1963 The cellular basis of morphogenesis and sea urchin development. Int. Rev. Cytol., 15: 139–214.

Holtfreter, J. 1931 Über die aufzucht isolierter Teile des Amphibienkeimes. II. Züchtung von Keimen und Keimtellen in Salzlösung. Roux' Arch., 124: 404–466.

Hotchkiss, R. D. 1948 A Microchemical Reaction Resulting in the Staining of Polysaccaride Structures in Fixed Tissue Preparations. Arch. Biochem., 16: 131–141.

Hyman, L. H. 1951 The invertebrates, Vol. 2. New York: McGraw-Hill, 1951.

Keller, V. 1894 Die ungeschlechtlich Fortpflanzung der Süsswasser-Turbellarien. Jen. Zeit. Naturw., 28: 370–407.

Johnson, W. H., C. A. Miller and J. H. Brumbaugh 1962 Induced loss of pigment in planarians. Phys. Zool., 35: 18–26.

Le Moigne, A., M. J. Sauzin, T. Lender et R. Delanault 1965 Quelques aspects des ultra-structures du blasteme de régénération et des tissus voisins chez Dugesia gonocephala (Turbellarié, Triclade). C. R. Soc. Biol., 159: 530–534.

Lender, Th. 1962 Factors in morphogenesis of regenerating fresh-water planaria. Advances in morphogenesis, Vol. 2 (ed: M. Abercrombie and J. Brachet): 305–331. New York and London: Academic Press, 1962.

Lender, Th., and A. Gabriel 1960 Etude histochemique des néoblastes de Dugesia lugubris (Turbellarié Triclade) avant et pendant la régénération. Bull. Soc. Zool. Fr., 85: 100–110.

Lillie, R. D. 1944 Various oil soluble dyes as fat stains in the supersaturated isopropanol technique. Stain Tech., 19: 55.

—— 1954 Histopathologic technic and practical histochemistry. New York: Blakiston and Co., 1954.

Lindh, N. O. 1957 Histological aspects of regeneration in Euplanaria polychroa. Ark. Zool., 11: 89–103.

Murray, M. R. 1927 The cultivation of planarian tissues in vitro. J. Exp. Zool., 47: 467–505.

—— 1928 The calcium-potassium ratio in culture media for Planaria dorotocephala. Phys. Zool., 1928: 137–146.

—— 1931 In vitro studies of planarian parenchyma. Arch. f. Exper. Zellforsch., 11: 656–668.

Pedersen, K. J. 1959 Cytological studies on the planarian neoblast. Zeitsch. f. Zellforsch., 50: 799–817.

—— 1961a Studies on the nature of planarian connective tissue. Zeitsch. f. Zellforsch., 53: 569–608.

—— 1961b Some observations on the fine structure of planarian protonephridia and gastrodermal phagocytes. Zeitsch. f. Zellforsch., 53: 609–626.

Prenant, M. 1922 Recherches sur le parenchyme des Plathelminthes. Arch. Morph. gén. exp., 5: 1–72.

Reisinger, E. 1959 Anormogenetische und parasitogene Syncytienbildung bei Turbellarien. Protoplasma, 50: 627–643.

Seilern-Aspang, F. 1957 Polyembryonie in der Entwicklung von *Planaria torva* (M. Schultz) auf Deckglasskultur. Zool. Anz., *159:* 193–202.
——— 1958 Zum Problem von Organisationsvorgangen bei dreidimentionalen Wachstum in Gewebekultur. Zool. Anz., *160:* 1–7.
——— 1960 Beobachtungen über Zellwanderungen bei Tricladengewebe in der Gewebekultur. Roux' Arch., *152:* 35–42.
Skaer, R. J. 1961 Some aspects of the cytology of *Polycelis nigra*. Quart. J. Micro. Sci., *102:* 295–317.
Stephan-Dubois, Fr. 1961 Les cellules de régénération chez la Planaire *Dendrocoelum lacteum*. Bull. Soc. Zool. Fr., *86:* 172–185.

Török, L. J., and P. Röhlich 1959 Contribution to the fine structure of the epidermis of *Dugesia lugubris O. Schm.* Acta biol. Sci. hun., *10:* 23–48.
Wolff, Et. 1961 Cell migration in planaria. Exp. Cell Res., Suppl., *8:* 246–259.
——— 1962 Recent researches on the regeneration of planaria. Regeneration (ed: D. Rudnick): 53–84. New York: The Ronald Press Company, 1962.
Wolff, Et., and Fr. Dubois 1948 Sur la migration des cellules de régénération chez les Planaires. Rev. Suisse Zool., *55:* 218–227.

PLATES

PLATE 1

All stained cells (4, 5, 6 and 8) were fixed with Zenker's fluid on a cover glass and then stained by the specified method. All live cells were photographed through an inverted microscope.

4　Spherical neoblasts without processes released from the surface of a twenty-two and one-half hour old fragment in saline solution. Note the scant cytoplasm. The nucleus of a neoblast usually contains one nucleolus of about 0.8 μ in diameter, but the nucleoli of most of the nuclei in the picture are out of the focused plane and not visible. Two nuclei in the left contain a few nucleoli. The vacant spaces of the size of nucleoli seen in the cytoplasm of a few neoblasts are the sites of the lipid droplets which were extracted during the preparation of the specimen. Azure A-eosin B stain at pH 5.0. $\times$ 1150.

5　Round neoblasts without processes released from the surface of a freshly minced fragment in saline solution. Methyl green-pyronin Y stain.

6　Same but stained after 60°C 1 N HCl treatment for 15 minutes. The cytoplasm and the nucleolus no longer take pyronin Y.

7　When a mixed cell suspension is allowed to settle in a culture dish, only the neoblasts start to aggregate. Due to their unique mechanism of locomotion described in the text, the aggregation is achieved locally between nearby neoblasts. Other cell types remain separate. The round cells present in the picture are too far out of the focus to be identified since they are not attached to the bottom as the neoblasts and the muscle cells are. Live, in 3% calf serum-containing saline solution. ne, neoblasts; mus, muscle cells. $\times$ 281.

8　One of the typical neoblasts with a primary process (p) and a secondary process (s) isolated in the basal medium. Azure A-eosin B stain at pH 5.0. $\times$ 1200.

9–10　Typical neoblasts with a primary process. Live, in the basal medium.

11–16　The development and the alteration of the processes of isolated neoblasts. The numerical figures in the pictures indicate the time elapsed from figure 11 in hours, minutes and seconds. The neoblast b rotates clockwise and one of its processes attaches to the tip of the primary process of another neoblast a (figs. 11–12). The neoblast b then develops a long process. The neoblast a has a typical bipolar form in figure 11, but it becomes unipolar when the tip of its primary process attaches to the neoblast b (fig. 13). Then, becoming bipolar again (fig. 14), it rotates its cell body (fig. 15). Then, the former secondary process becomes longer and the former primary process is shortened (fig. 16). The processes show feeble appearance and are motionless during the rotation of the cell body. The neoblast c remains somewhat inactive throughout, but it develops its secondary process. Live, in the basal medium.

PLATE 1

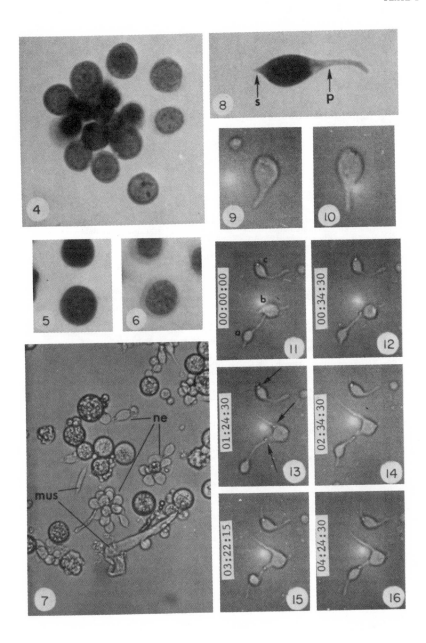

PLATE 2

EXPLANATION OF FIGURES

All cells illustrated here were live and photographed through an inverted microscope.

17–24   The movement of lipid droplets and the rotation of the cell body in an isolated unipolar neoblast are shown. A large arrow points the site of the secretion where an accumulation of lipid is also seen. Smaller arrows point lipid droplets in the process (fig. 17). After some amount of go-and-back movement, the lipid droplets accumulate at the end of the process (fig. 24). The rotation of the cell body is seen in figures 17–21. The process appears feeble during the rotation of the cell body (fig. 18). Cultured in 3% calf serum-containing saline solution.

25   Neoblasts in monolayer in saline solution. The photograph was taken at the focus of the middle plane of the cells. The lipid droplets are clearly demonstrated as black regions. Local aggregations have started (arrows). × 154.

26   A large aggregate of neoblasts in the middle is the result of the union of small aggregates. The surrounding neoblasts will eventually join the main aggregate. Photographed at the focus of the median plane of the cells. Cultured in the basal medium. × 127.

27   The aggregation of neoblasts starts with several local aggregations and then these aggregates unite into a larger aggregate. Aggregates a, b, c, d and possibly e will unite. Aggregates f and g will unite into one but may not join together with the others because the distance between f and g, and the others is farther than the reach of their processes. Cultured in the basal medium. × 127.

28–33   The process of the aggregation of neoblasts. The initial stage of the aggregation is seen at the upper left of figures 28–29. The tips of the primary processes of neoblasts attach to each other. The local aggregations first occurred (figs. 28–30) and then two of them united (fig. 32). Finally they became a large spherical aggregate (fig. 33). A muscle cell contained among the isolated neoblasts (arrow in fig. 28) soon collapsed (arrow in fig. 29). It underwent degeneration and finally cytolysed (arrow in fig. 32). The mitosis of a neoblast is indicated by the arrow in figures 30–31. Cultured in saline solution containing 3% calf serum.

34–35   Spherical aggregates of neoblasts floating in saline solution.

28

PLATE 2

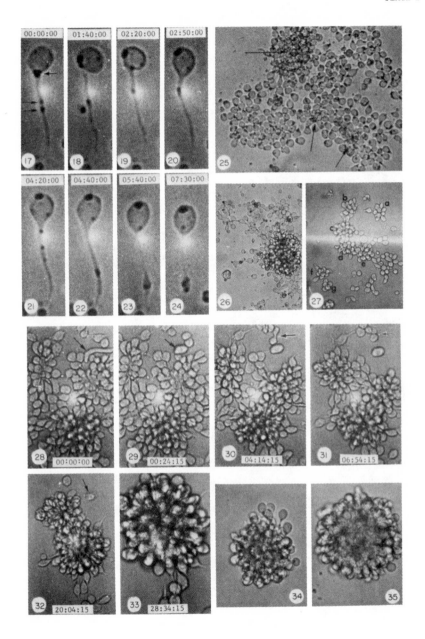

29

PLATE 3

EXPLANATION OF FIGURES

All cells were fixed with Zenker's fluid and stained with Azure A-Eosin B at pH 5.0.

36 Neoblasts in the parenchyma of an intact animal. The processes of the neoblasts are apparent. They are strongly basophilic as well as the perikaryons. Arrows point some processes of the neoblasts. Paraffin section. × 780.

37 Neoblasts from the surface of a three-day-old fragment. The fragment was cultured in saline solution for the first two and one-half days and in the basal medium for the last half day. Fixed on a cover glass. × 780.

38 Aggregating neoblasts from the same culture shown in figure 37. Fixed on a cover glass. × 830.

39 A small group of aggregating neoblasts from the same culture shown in figures 37–38. Note the orientation of the processes. Fixed on a cover glass. × 830.

40 Aggregating neoblasts in saline solution (in 3-day-old culture) were pulled from the center of the aggregation and stretched by the surface tension when the medium was withdrawn. Fixed on a cover glass. × 753.

PLATE 3

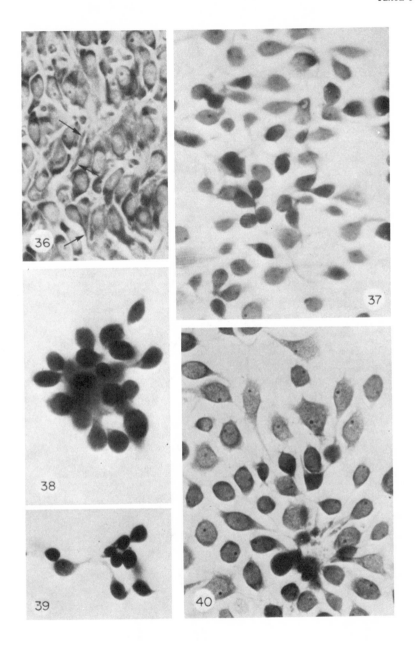

41–49   The interaction between a fixed parenchyma cell and aggregating neoblasts is shown in several frames of the time-lapse cinematograph taken through an inverted microscope. A single fixed parenchyma cell moved through the middle of the aggregating neoblasts from one side to the other by pushing them apart. Then, the fixed parenchyma cell with its filopodia penetrating between the neoblasts prevented the neoblasts from making a tight aggregate. In the upper left of figure 41, a part of the fixed parenchyma cell with one of its filopodia is seen. In figure 42 the fixed parenchyma cell enters into the aggregating neoblasts; only the end of one of its filopodia is seen. In figure 43 the fixed parenchyma cell is completely inside of the aggregating neoblasts. The aggregating mass of neoblasts expands, indicating the forcible passage of the fixed parenchyma cell. In figure 44 a part of the sheet-like filopodia of the fixed parenchyma cell starts showing at the opposite side of the aggregating neoblasts. In figure 45 the cell body with one of its filopodia starts showing at the upper end of the aggregating neoblasts but another filopodium penetrates through the aggregating neoblasts and appears at the opposite side. The fixed parenchyma cell apparently prevents one of the neoblasts (marked by X) from aggregating (figs. 44–49). The initial stages of aggregation of the neoblasts are seen in the lower middle of figures 41–42. Note the extended primary processes of the neoblasts. The small black arrows show the filopodia of the fixed parenchyma cell and the large black arrows trimmed by white show the cell body. Live, in saline solution containing 3% calf serum.

50   A fixed parenchyma cell and a neoblast separated from a three-day-old fragment. The nucleus of the fixed parenchyma cell is much smaller than that of the neoblast. Dark granules in the cytoplasm of the fixed parenchyma cell are lipid droplets. The fragment was cultured in saline solution for the first two days and in the basal medium for one day. Fixed in 2% osmium tetroxide (pH 7.4) for 30 minutes. Phase contrast.

32

PLATE 4

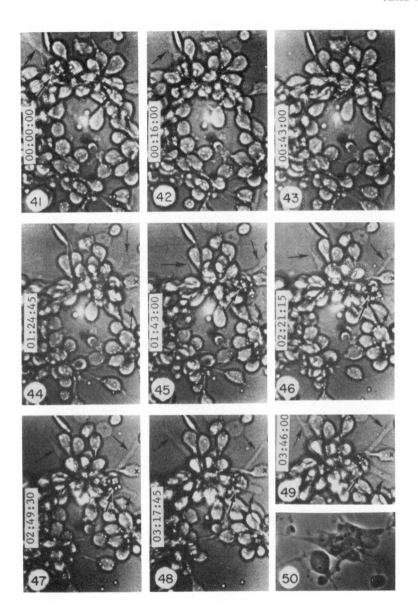

PLATE 5

EXPLANATION OF FIGURES

51–56 The active movement of a gastrodermal cell freed by the trypsin treatment was photographed at five-second intervals. Live, in saline solution. Phase contrast.

57–58 A fragment treated with trypsin was flattened under a cover glass and then photographed. Figure 57 shows pseudopodia of the gastrodermal cells. Figure 58 shows one of the multinucleated gastrodermal cells. The cell also contains a cell with scanty cytoplasm in one of its vacuoles (large arrow). Live, in saline solution. Phase contrast.

59 A fragment was fixed immediately after having been minced in saline solution and then paraffin sectioned. The fragment contains all sorts of tissues. Formalin-alcohol-acetic acid fixation. Delafield's haematoxylin-eosin stain. × 106.

60 A twenty-two and one-half hour old fragment cultured in saline solution was sectioned. The gastrodermis (gt) has united into a continuous mass and covers the entire fragment. They are extremely vacuolated. In the periphery of the inner mass, a considerable accumulation of neoblasts is noted. Necrotic figures of other cells in peripheral area are prominent. Formalin-alcohol-acetic acid fixation. Paraffin section. Delafield's haematoxylin-eosin stain. × 106.

34

PLATE 5

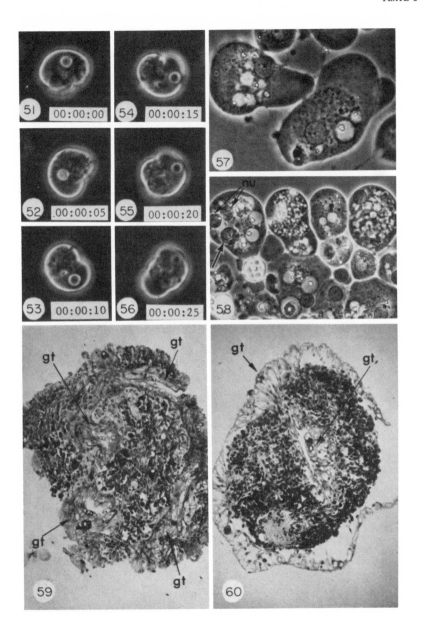

PLATE 6

EXPLANATION OF FIGURES

61   A middle section of a fifty and one-half hour old fragment cultured
in saline solution. The gastrodermal cells (gt) have moved out of
the fragment and completely separated themselves from the rest. The
remainder of the fragment demonstrates an increased density of neoblasts. Mitotic figures of neoblasts (arrow) are seen. Formalin-alcohol-acetic acid fixation. Delafield's haematoxylin-eosin stain. Paraffin
section. × 271.

62   A middle section of another fifty and one-half hour old fragment cultured in saline solution. The separation of tissues is noted. Upper
left: parenchymal tissue (pa), mostly neoblasts; upper right: gastrodermal tissue (gt); lower right: epithelial tissue (ep); lower left:
nervous tissue (nv). Formalin-alcohol-acetic acid fixation. Delafield's
haematoxylin-eosin stain. Paraffin section. × 271.

PLATE 6

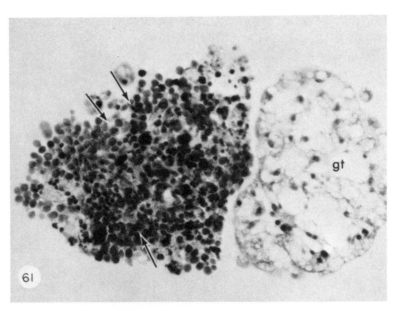

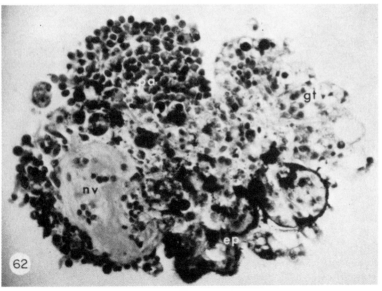

37

The Cellular Mechanism of the Formation of a Regeneration Blastema of Fresh-water Planaria, *Dugesia dorotocephala*

I. THE BEHAVIOR OF CELLS IN A TINY BODY FRAGMENT ISOLATED *IN VITRO*

TEIICHI BETCHAKU

The existence of particular parenchyma cells as occupying the major part of the early regeneration blastema of fresh-water planarians has been described by many authors since Keller (1894) (see Brøndsted, '55, for a review). These cells, identified by their oval or pear shape, scanty cytoplasm, and strong basophilia, are now generally called "neoblasts" after Buchanan ('33) and Wolff and Dubois ('48). The basophilia of the neoblasts has been found to be due to the high content of RNA in the cytoplasm (Pedersen, '59).

The mechanism of their accumulation at the wound, as well as their function, has been a major topic of investigation in recent decades. A group of French workers (Wolff and Dubois, '48; Dubois, '49; Lender, '52; Lender and Gabriel, '60; Stephane-Dubois, '61; Fedecka-Bruner, '65) argue that *neoblasts* are embryonic totipotent cells which somehow remain in an undifferentiated state in the parenchyma. At the site of a wound, such cells migrate from throughout the body to accumulate as a regeneration blastema which then differentiates into every part of the new regenerate. The evidence for this concept of extensive migration of neoblasts over considerable distances was first supplied by the x-ray irradiation technique (Wolff and Dubois, '48; Dubois, '49). According to these authors, the inability of irradiated animals to regenerate was considered due to the selective destruction of neoblasts by a controlled dose of x-rays (5,000 r units). The irradiated animals regenerate, they argued, only when fresh neoblasts migrate to the wound either from the unirradiated area of the body or from a grafted unirradiated tissue fragment. Additional evidence of migration was supplied by counting the number of neoblasts at different body levels from the wound during the course of regeneration (Lender and Gabriel, '60; Stephane-Dubois, '61).

The x-ray irradiation experiments of Wolff and Dubois did not, however, actually tell whether neoblasts migrated alone to the wound or together with other types of cells, nor how neoblasts migrated through parenchyma. Hay, in a recent critique ('66), mentions the possibility that migrating cells revived the old irradiated tissues as reported to occur in amphibian limb regeneration (Polezhaev et al., '60, '61). The possibility that neoblasts derived from differentiated tissues was also not ruled out. The time-differential frequency gradient of neoblasts during regeneration established by the counting technique of Lender, Gabriel and Stephane-Dubois may not necessarily support the concept of directed migration neoblasts. The large portion of the interior of this animal is occupied by the highly branched gastro-vascular system. The narrow spaces between these intestinal branches are filled with soft and flexible parenchyma in which neoblasts are distributed. Very little attention has been given to the possibility that general morphogenetic movements of cells and tissues other than that of neoblasts can occur during the course of regeneration. Thus, there is a possibility that the count of neoblasts may simply represent the volumetric changes of the parenchyma at various levels from the wound. If the migration of neoblasts through parenchyma (neoblasts are parenchymal cells) is the issue, the density of neoblasts in parenchyma, not the mere number of neoblasts per body level, should have been considered. The behavioral characteristics of neoblasts and fixed parenchyma cells revealed in *in vitro* culture (Betchaku, '67) contradict the idea of the extensive free migration of neoblasts through parenchyma and favor the local origin of neoblasts of the blastema.

Not only has the concept of neoblasts as highly migratory cells not been satisfactorily proved, but it imposes very difficult questions. For example, what provokes the migration of neoblasts and what guides them to the site of the wound through the parenchyma? Wolff ('62) proposed two hypotheses: the emission of stimulatory substances by injury, which diffuse along the body and the stimulation conducted by the nervous system. Wolff, however, noted, "The initiation of neoblasts activity could be produced by an injury such as a scalpel incision. It is not necessary that the planarian be amputated or that the regeneration of an important organ be involved. But it is necessary that the wound be open. A serious lesion such as a burn, an electrocoagulation, and irradiation by x-ray never provokes a stimulation of migratory cells."

While the wound must be open for regeneration to occur, the inhibition of regeneration by the coalescence of an open wound has long been known. Goetsh ('21) first reported such inhibition and proposed the counteraction between coalescence or "Verwachsung" of the wound and regeneration. Li ('28), Okada and Sugino ('34, '37), Okada and Kido ('43), Brøndsted ('39) and Brøndsted and Brøndsted ('54) concluded, after almost every possible combination of planarian body grafts, that the wound coalescence is not the only requirement for the inhibition of regeneration. Internal structures and the body level of united pieces, if reversed, must be matched. They considered that this phenomenon was due to the antagonism of polarities. However, the experiment of Sugino ('36), described below, shows clearly that coalescence of the wound is nonetheless the main factor in prohibiting the creation of new tissue. If the body levels of each graft are matched, the internal structures should also be closely matched, thus making coalescence of severed tissues much faster. Sugino decapitated the animal by a v-shape incision in the prepharyngeal area and filled the wound with a matching triangular piece cut from the lateral edge of the discarded anterior portion of the body. If fits were perfect, no internal structure was exposed and no new tissue appeared. There was no head regeneration among these animals.

The rate of regeneration of a new head, measured by the appearance of eye spots, at each level of the body is more or less fixed by the time-graded regeneration field (Brøndsted, '56). What would happen if a number of open wounds of large size are made simultaneously at various places of the body? Under such circumstances, involving stimuli from many wounds or from the nervous system as proposed by Wolff ('62), would activation, guidance, and migration of neoblasts be confused? Or, would regeneration at each wound be retarded because the neoblasts available would be divided among the wounds? Brøndsted and Brøndsted ('54) did just that experiment to test this by measuring the rate of head regeneration of *Dendrocoelum lacteum*, *Euplanaria* (*Dugesia*) *lugubris* and *Polycelis nigra* in segments of varying size but decapitated at the same level. The pieces varied from a decapitated full size animal to a cross-piece of about one-eighth of the body length. With *Bellocephala punctata*, they compared the rate of head regeneration between animals with only an anterior cut and animals with large lateral incisions in addition to the anterior cut. All regenerated at the same rate, and they concluded that neither the size of the regenerating piece nor the extent of wounding influences the rate of regeneration of eyes at a given level of the body. Thus, it appears that, the mechanism of regeneration should be sought locally.

Most body levels of *Dugesia dorotocephala*, except for the extreme anterior and posterior portions, are able to form a blastema and regenerate either a head or a tail depending on which end of the body has been lost (Child, '11; Sivicks, '32). In other words, every minute portion of this animal has the potential to become either an anterior or a posterior structure. Since a blastema formed on either anterior end or posterior end, and since at a given level the density of neoblasts is identical, control of morphogenesis of the blastema must lie in the remaining portion of the body. Furthermore, if the neoblasts are supplied locally they are presumably identical. The decisive polarity of the remaining portion of the body can still be recognized in the regeneration of a very small isolated piece of the body. Child ('11, with *Dugesia dorotocephala*) reported normal regeneration from prepharyngeal cross-pieces of one sixteenth of the body length or about 1 mm. Morgan ('01, with *Dugesia lugubris*) observed normal regeneration even from a lateral half of a short cross-piece such as used by Child. The minimum length of the cross-piece for the preservation of original polarity and thus capable of regen-erating normally, was calculated to be about 1 mm in the prepharyngeal area by Okada and Kido ('43) in a similar species, *Dugesia gonocephala*.

These observations reveal that material sufficient for both anterior and posterior regeneration blastema, or even for an additional lateral one, can be supplied locally under a regulatory control in a body fragment as small as 0.5 mm³. If the blastema is composed of parenchymal neoblasts alone, there must be a sufficient number of neoblasts in every portion of the body to meet the demand or they can be proliferated. Active proliferation of neoblasts during the course of blastema formation has been reported by Lindh ('57) and Pedersen ('58). If the original neoblasts and those resulting from their proliferation are not sufficient to form a blastema, the dedifferentiation of differentiated cells into *neoblasts* must also be considered as occurring. It is also possible that, regardless of the number of available parenchymal neoblasts, dedifferentiation of differentiated cells into *neoblasts* will contribute to the formation of a regeneration blastema. These questions must also be answered.

The present investigation will deal with the behavioral patterns of cells and tissues of fresh-water planarians and with their interactions in a small isolated cellular community in order to understand the mechanism of the formation of a regeneration blastema, on the assumption that interactions of local tissues play an important role.

MATERIALS AND METHODS

A cross-piece was first cut out midway between the head and the pharynx of asexual *Dugesia dorotocephala* 16 mm to 20 mm long in saline solution (Betchaku, '67: NaCl 2,188 mg; KCl 31 mg; CaCl₂ 63 mg; NaHCO₃ 125 mg; Neomycin sulfate 100 mg; H₂O 1,000 ml; pH 7.3). The pieces were then sliced dorso-ventrally into small tissue cubes of 0.1 mm³ to 0.03 mm³ on a wax block (fig. 1). They were kept undisturbed in saline solution at 17°–18°C. At intervals the tissue cubes were pipetted into fixative (Formalin 10 ml; 95% Ethanol 40 ml; Glacial acetic acid 2 ml). They were embedded in paraffin, sectioned and stained with Delafield's haematoxylineosin.

40

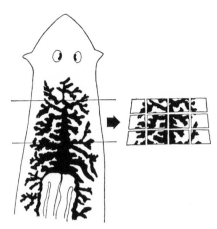

Fig. 1 A small cross-piece was cut out from the prepharyngeal area, and sliced further into smaller cubes for the experiment. The distribution of the gastrovascular system, visualized by feeding the animal with carmine mixed with egg yolk, is shown in black.

Animals were cleaned in several changes of sterile spring water containing 0.02% Neomycin sulfate for 24 hours before the operation.

Terminology

The terms, "neoblasts" and "fixed parenchyma cells", are used in this report as in the previous report (Betchaku, '67) for the sake of identifying the two cell types which together compose planarian parenchyma. One has an oval or pear-like shape with scant but strongly basophilic cytoplasm and the other has highly attenuated slightly acidophilic cytoplasm and comprises the general background of the parenchyma. Prenant ('22) described the former type as "*cellules libres du parenchyme*" which, he thought, wander about the parenchymal *syncytium* represented by the cells of the latter type which he called "*cellules fixes du parenchyme*." The *cellules libres* were believed by many authors to be the embryonic totipotent cellular stock for regeneration. Accordingly, these cells were called by such names as "Stammzellen (Keller, 1894)," "formative cells (Curtis, '02," "Regenerationszellen (Bartsch, '23)," and "neoblasts (Buchanan, '33; Wolff and Du-

bois, '48)" which imply their assumed function. The term *cellules fixes du parenchyme* was translated into English as "fixed parenchyma cells" by Pedersen ('61) who, incidentally, disproved the long standing concept of parenchymal syncytium through electron microscopical studies. In spite of much work on the problem, however, the nature and function of these parenchymal cells are yet to be learned. The words "neo-" and "fixed" have little meaning in the present report other than in identifying two types of parenchymal cells. For a detailed description of these cell types, see Pedersen ('59, '61) and Betchaku ('67).

OBSERVATIONS

Because of the unique body plan of this animal, each body fragment contains almost equal varieties of cells and tissues but differing slightly in configuration. Most of them have a cubic form with the dorsal and ventral sides covered with epidermis and four cut surfaces. Some cubes became broken apart in the middle because of the gastro-vascular cavity and thus had only one side (either dorsal or ventral) covered with epidermis and five open sides. The pieces containing a part of the original lateral ridge have five faces with dorsal and ventral epidermis and three open sides. The open sides of these body fragments may be lined with gastrodermis or may have bare parenchyma exposed to the culture medium (fig. 1). Figures 2–3 are two samples of these isolated body fragments.

For the first few hours, the fragments remain essentially motionless with slight occasional undulation and hold the original cubic form. The surface of the exposed parenchyma and gastrodermis stays smooth, and cavities in or protrusions of the interior tissues are not observed.

Behavior of gastrodermis

Planarian gastrodermis has been described as consisting of two types of cells, columnar phagocytic cells and small, club-shaped cells filled with small granules (Minot, 1877; von Graff, '12–'17; Hyman, '51). The latter type has been postulated to serve as gland cells or protein reserves (Arnold, '09; Ijima, 1884; von Graff, 1899; Willier et al., '25). The differences in cell

41

shape and character of these two types, however, disappear when the cells are dissociated by trypsin or are in a dissociated body fragment (figs. 3–10; see also Betchaku, '67). Although further work will be necessary before a definite conclusion on their nature can be drawn, the gastrodermal cells will be treated as a single type of cells in this report.

After five to ten hours, the most noticeable movement of cells is the migration outward of gastrodermis from the open sides of the fragment (figs. 7–8). Most of the gastrodermis becomes loosened from underlying parenchymal tissues and is involved in the outward movement. The severed pieces of gastrodermis in the fragment fuse into a continuous mass. The gastrodermis tends to spread over the surface of the fragment. Contact with the parenchymal surface occurs only at irregular intervals (fig. 9). Alternatively, it grows into the medium without adhering to any particular substrate (figs. 10, 36). Gastrodermal contacts are made mostly between parenchymal cells or muscle cells. Whether these gastro-muscular and gastro-parenchymal contacts are the original contacts or are newly formed is uncertain. Gastrodermis *in situ* is lined with either intestinal muscles or fixed parenchymal cells described as *Plasmabrücken* by von Graff ('12–'17). Gastrodermal cells and neoblasts seem to be the least cohesive to each other. No attachment of the migrating gastrodermis is made on the parenchymal surface where neoblasts are densely accumulated.

The rate of gastrodermal expansion is so rapid compared with the movement of other tissues that, within a day, gastrodermis is the dominant tissue on the surface of the fragment (figs. 9–10). The volumetric increase of gastrodermis is achieved by both the fusion of severed gastrodermis and the enormous vacuolation of the gastrodermal cells rather than by their proliferation. The supervacuolation of the migrated gastrodermis is very likely caused by intercellular adhesiveness that stretches the cells to an extreme enlargement: if these supervacuolated cells of a 22.5 hour old fragment (figs. 9–10) are broken down into individual cells artificially, each dissociated cell immediately re-

sumes its regular much less vacuolated form (fig. 4), morphologically similar to that of freshly dissociated cells (after trypsin treatment) from an intact animal (fig. 5). In other words, this excessive vacuolation only occurs in cells arranged in sheets or groups and not in single cells cultured in the same medium. The mechanism of this supervacuolation has yet to be studied, but it should not be mistaken for a fixation artifact. Fragility of supervacuolated gastrodermis, as previously reported (Betchaku, '67), may not necessarily be due to the low adhesiveness between the cells but may be due to the structural weakness of cytoplasm stretched extremely thin.

The volumetric expansion of gastrodermis is undoubtedly the major motive force of its spreading. The gastrodermis may, therefore, drag the underlying parenchymal tissue to elongate with it by its original contacts. This must cause the initiation of frequent structural inversions of isolated fragments (figs. 9–11, 29, 36).

The differential adhesiveness of gastrodermis to other types of cells has another significance. The gastrodermis shows more affinity to the epidermis than to any other tissues. It adheres to the outersurface of the epidermis, not only non-ciliated dorsal epidermis but also ciliated ventral epidermis as well, so that very often gastrodermis crawls over the non spreading epidermis (see *Behavior of epidermis*) when it comes into contact with the latter on the surface of the fragment (figs. 12–20). Gastrodermis stays firmly on the surface of the epidermis even on the ciliated ventral epidermis, probably because of the gastrodermis' strong affinity to the non-ciliated lateral sides of the epidermis. Histological examination of a fixed specimen reveals that the gastrodermis is tightly attached in every contact area with epidermis. The effect of this differential affinity of the gastrodermis is well demonstrated in figures 19–20. The migrated gastrodermis is always thicker on the epidermis and thinner over the parenchymal tissue when such alternative substrates are given to the spreading gastrodermis. It is often extremely stretched without touching the parenchyma.

In situ, each gastrodermal cell is columnar shaped and its structure is polarized with its nucleus located proximal to the base. When supervacuolized, the cell seems to become depolarized as seen in many of the illustrations of this paper. It is of interest, however, to note that such supervacuolized gastrodermis does devacuolize itself and resumes its original columnar shape on the surface of the epidermis to which the gastrodermis manifests positive adhesion. In figure 30, the linearly arranged narrow feet of less vacuolated gastrodermal cells may be noted against the outersurface of a spherical aggregate of epidermis.

The condition and the mechanism for the gastrodermis to resume its regular morphology is to be studied further. However, it is certain that its high adhesiveness to the substrate is necessary for gastrodermal morphogenesis.

Behavior of epidermis

The epidermis consists of a single layered cuboidal epithelial cells which are somewhat taller on the dorsal side of the body than on the ventral side. Many rhabdites are present in the epidermis. The epidermis is lined with basement membrane which appears 0.2–0.5 μ thick under a light microscope. Pedersen ('61) described the basement membrane of *Dugesia tigrina* as about 1 μ thick and consisting mostly of filaments of 75–100 Å in diameter by electron microscopy. Only the ventral epidermis is ciliated. A singly isolated ciliated ventral epidermal cell is illustrated in figure 6. The localized ciliation is best observed when a very thin cross section of the living body, in saline solution, is examined perpendicularly through a phase contrast microscope. Identification of dorsal or ventral epidermis in a histological section of an isolated body fragment can also be made with regard to the underlying tissues: the subepidermal pigment is much denser dorsally; subepidermal musculature is much thicker ventrally; the central nervous system, namely two longitudinal nerve cords and numerous transverse commisures, are located ventrally just below the ventral muscle layers.

The behavior of epidermis is characteristically different from that of gastrodermis. Epidermis is neither able to roll over the bare surface of parenchyma nor able to grow out freely into the medium without substrate as gastrodermis does. The free edge of the epidermis of the fragment stays in its original position without advancing. Contrary to the general notion (see DISCUSSION), some flattening of individual cells within the epidermal sheet between the stationary free edges occurs after five to ten hours in culture (fig. 7–8). The epidermis, because of the increase in dimension, either curves under to engulf the underlying tissues (fig. 8), loops away from its basement membrane and later clumps into many folds (fig. 21), or accumulates at a non-advancing front (figs. 34–35).

As described earlier, the difference of the rate and the nature of the expansion between these two opposing layers across the parenchymal complex is such that, if one side of the fragment is lined with gastrodermis, the gastrodermal side expands faster and inverts the whole structure of the fragment (figs. 9–11, 29, 36). Or, a detached gastrodermis alone simply migrates outward to cover the entire surface of the fragment even over the epidermis (figs. 19–20).

Epidermis is the most resistant tissue to culture conditions and is the only tissue which does not split or break during the entire period of culture. When epidermis does not spread over the parenchyma, that is, if there is no substrate with sufficient adhesiveness to the epidermis, individual epidermal cells increase their mutual contact, and aggregate.

The outersurface of epidermis, lacking adhesiveness to other cellular material except for the migrated gastrodermis, is always left free. This creates a cavity in the center of the inversely aggregated epidermis (figs. 9–11). When they come into contact, the lateral and proximal sides of the epidermal cells fuse with each other to form a syncytium-like structure (figs. 10–11, 30–31, 34–35).

A variety of epidermal aggregates are formed on the surface of the fragment when inversion does not occur. If the initial piece of the epidermis is large compared with the amount of other tissues, especially of gastrodermis, the epidermis curves under, possibly by its mutual cohesion which tends to shorten its base, displacing most tissues beneath it to form an epithelial sphere (figs. 20–27, 30–31). There is a specific order in the movement of tissues from such clumping epidermis. The first cells to move out are the gastrodermal cells. The gastrodermal cells form a small aggregate of their own and usually float away because of their low affinity for all tissues except epidermis. Following the gastrodermis are the parenchymal elements. The cellular elements which often remain inside the epidermal sphere are the cells which structurally attach to the epidermis, such as muscles and nervous plexi innervating both subepidermal muscles and epidermis. The muscle cells and diverged nerve fibers, some in process of degeneration, then loosely fill the inner space of the epidermal sphere together with some rhabdite-forming cells and scattered granules of subepidermal pigments (figs. 22–27, 32). The rhabdite-forming cells in the epidermis usually remain in position between epithelial cells.

Within 32.5 hours in culture, the parenchymal cells surrounding subepidermal muscles and subepidermal nervous plexi move out to join the main mass of the parenchyma and leave vacant spaces at their previous sites. These vacant spaces between the remaining muscles and nerve plexi indicates that parenchymal cells migrate by themselves and are not pushed out by the adhering epidermis (figs. 23–25). Some subepidermal pigments remain in most of the epidermal aggregates. Most of these granular pigments, however, move out of their original subepidermal area with other parenchymal cells, suggesting the possibility of "parenchymal movement" *in vivo* (Betchaku, '67).

The subepidermal muscles and nerve plexi are displaced from an epidermal sphere only when a clumping epidermis is relatively small and the syncytium-like epidermal aggregate fills the inside of the sphere (figs. 30–31). Otherwise, these muscles and nerve plexi remain with a space around them (figs. 23–25, 32). This indicates that, when separated from the sphere, they are actually squeezed out, in a relative sense, by the cohesive force of aggregating epidermal cells. These highly differentiated cells such as muscles and nerves are neither mobile nor do

they dedifferentiate into mobile forms of any kind during the course of tissue segregation. If the fragment is large with a large amount of epidermis, the flattened epidermis conceals the fragment trapping many tissues inside, possibly by a process similar to normal regeneration after the intact muscles around the wound contract to narrow the opening. Thus, variations of epidermal aggregates have been observed consisting of from almost a pure aggregate of thickened epidermal cells with only a few rhabdite-forming cells, to a large heterogeneous cellular mass enclosed by a flattened epidermis (figs. 22–27, 30–33).

Behavior of parenchyma

The interior of an intact planarian body between organs is filled with both fixed parenchyma cells and neoblasts. There is very little intercellular space, and what there is has a width of only a few hundred Å (Pederson, '61, with *Dugesia tigrina*). Unicellular gland cells of various types are quite abundant and scattered in the parenchyma. In the isolated body fragment, severed branches of gastrodermis start to unfold and fuse into a continuous mass and segregatae themselves externally (figs. 7–10). The epidermis aggregates either internally or externally as described earlier.

Within the first 32.5 hours, the parenchymal cells migrate from the narrow spaces between organs and tissues to aggregate themselves into a condensed parenchymal mass (figs. 22–25). They display no spreading tendency, revealing a strong mutual affinity with little affinity for other tissues. Regular unicellular inclusions of the parenchyma are found in this parenchymal condensate. In the second day of culture, when the separation of tissues in the fragment becomes evident, the frequency of neoblasts on the exposed surface of parenchyma increases. In a small parenchymal condensate, isolated in saline solution, the accumulation of neoblasts reaches its maximum being almost devoid of fixed parenchyma cells and other unicellular inclusions of parenchyma (figs. 26–27).

Figures 26 and 27 are two different sections from the same 50.5 hour old fragment in culture and show this difference of cellular susceptibility and mobility at work. Note two dense aggregates of neo-

44

blasts. At the center of the aggregate of neoblasts, the familiar radial arrangement of neoblasts is clearly seen (Betchaku, '67). Necrotic fixed parenchyma cells are observed between the segregated tissues. Unicellular gland cells of various types in the parenchymal mass follow the same fate as the fixed parenchyma cells. In the parenchyma of the fragments covered with the original epidermis (figs. 33–34), segregation between neoblasts and fixed parenchyma cells does not occur. In the parenchyma, partially covered tightly with either epidermis or gastrodermis, no segregation of parenchymal elements is observed (figs. 12–21).

Mitotic figures were observed only among the neoblasts and not among any other tissue. The mitotic activity of neoblasts, judged by the frequency of mitotic figures, was most abundant in the 2 to 3 day old fragments (figs. 28–29).

Mobility of tissues between conjugated body fragments

If two freshly cut-out fragments are placed side by side in light contact and kept undisturbed in saline solution, they will eventually fuse with each other. One such fused specimen, 22.5 hours old, is shown in figures 36–38.

Here too, the epidermis is not spread over the fragment, but is clumped together in various ways. One of the two fused fragments (the lower one of fig. 36) has been split open in the middle of the gastrovascular cavity. The structure of the dorsal half has been inverted. The dorsal epidermis has curled outward and aggregated into an inverted epidermal sphere with the normal configuration of its subepidermal structures (shown at higher magnification in figs. 10–11). The identification of either dorsal or ventral epidermis has been described earlier in this paper. The parenchyma, then, aggregates in the periphery surrounding the inverted epidermal sphere and the subepidermal structures. The gastrodermis detaches from most of the underlying parenchyma, becomes supervacuolated and stretched, and loosely surrounds the fragment. Conversely, the ventral half of the fragment has kept its tissue appearance, owing perhaps to its higher content of less motile

and less flexible tissues, e.g., nerve and muscle.

Whether there is any significant difference in the behavioral characteristics of dorsal epidermis and the ciliated ventral epidermis is not certain, for differences between dorsal and ventral epidermis seem to be due to the subepidermal tissues. In figure 36, the upper body fragment is undergoing the ordinary deformation of a singly isolated body fragment. This fragment is, judged by its tissue configuration (see *Behavior of epidermis*), a lateral piece which includes the original lateral edge of the body (pointed by an arrow in fig. 36). An interesting and probably one of the most significant features is that the contact between nervous tissues of each fragment seems to be the most solid attachment, although small in area.

DISCUSSION

Planarians are believed to be the lowest group of animals to have true mesoderm or entomesoderm. Being an acoelomate, the planaria has a definite and solid three layered body construction. All interior organs, including musculature and nervous system, are packed in the continuous coelomless parenchyma, which all together compose the inner layer lined with two epithelial layers; epidermis for the exterior and gastrodermis for the interior. According to the electron microscopical study of Pederson ('61, with *Dugesia tigrina*), only the narrow space of a few hundred Å with little intercellular material separates the cells in this inner layer, except for a somewhat widened space with densely packed filamentous material around the muscle cells. He reported that the real ground substance as observed in vertebrate connective tissue does not exist in freshwater planarians. This makes the inner layer of planarians an ideal composite cellular system for the study of cellular behavior in a heterogeneous population. Although the organogenesis of this animal which develops from an ectolecithal egg has not been well understood despite the pioneer works of Metschnikoff (1883), Ijima (1884) and Mattiesen ('04), this inner layer or the inner composite-tissue behaves, in general, as a unit with respect to the more homogenous epidermis and

45

gastrodermis. Only the parenchyma (fixed parenchyma cells and neoblasts), however, accounts for the mobility of the inner composite-tissue comparable to that of the two epithelia. In the present work morphological rearrangement or allocation of the tissue aggregates in the isolated body fragment was found to complete in two days. This was the result of the interactions of these three unit-tissues.

Although planarian epidermis and gastrodermis exhibit the characteristics of surface epithelia in their tendency to spread, the nature of their spreading is different. The spreading of a cell sheet requires an increase of the surface dimension. Planarian epidermis acquires this by flattening each epidermal cell in the sheet, and advances by the crawling mechanisms of the marginal cells. Because the "flattening" occurs even between non-advancing marginal cells, these two cellular activities should be regarded as two separate cellular movements which concur in the moving epithelial sheet.

As pointed out by Lewis ('26) and Thompson ('61), the form of cells in a soft tissue in situ is the expression of an equilibrium of several kinetic forces exerted on the cells, including the tension of the plasmalemma and the pressure exerted from the surrounding. The spherical form of freshly dissociated cells in suspension confirms it, and the flattened form of the detached epidermis should be regarded as a cellular expression of the newly established kinetic equilibrium. Without entering into a discussion of its mechanism, the detachment of epidermis from the basement membrane which creates another free surface certainly breaks the kinetic balance in the cell system, and the form of cells must be reformed accordingly. The uniformity of this thinning effect throughout detached epidermis supports this view (figs. 7 and 34).

The marginal cells of planarian epidermis are likely exercising a similar mechanism of migration to that of the cultured chick epithelial monolayer, namely "ruffled" membrane activity (Vaughan and Trinkans, '66). Thus, the spreading of planarian epidermis must be highly dependent on its adhesiveness to the substrate. Lash ('55, '56), working on skin regeneration of amphibian limbs, reported that epithelial cells do not hang together, but move individually. By contrast, planarian epidermis, in the case of the isolated body fragment advances without dissociating into individual cells. The reason for their inability to spread over the exposed surface of the parenchyma of the isolated planarian body fragment must be that there is inadequate "exudate" to serve as substrate for epithelial spreading (Weis, '59) over the open surface of the parenchyma. In normal planarian regeneration, the constriction of the wound area by muscle contraction may help the accumulation of such exudate besides narrowing the gap between the severed edges of the original epidermis.

The gastrodermis expands and spreads by supervacuolization of individual gastrodermal cells. The detachment of the gastrodermis from the underlying tissue may trigger the supervacuolization by altering the kinetic equilibrium exerted on the cells, but the involvement of adhesion between cells is apparent, because supervacuolization only occurs in the cells in the gastrodermal sheet and not in single cells. The resulting increase of the total volume from supervacuolization enables the gastrodermis to cover a huge area. It may adhere to the epidermis or the parenchyma, but it is not essential for the gastrodermis to have a substrate of proper adhesiveness to spread, although the degree of adhesion to the underlying tissue affects tightness of such gastrodermal coverage. No mitotic activity has been observed in either epidermis or gastrodermis during the observation, or for the first 50.5 hours after the isolation of a body fragment.

Contrary to either epidermis or gastrodermis, the parenchymal cells have no tendency to spread, but to accumulate. When either or both of the epithelia are removed, the parenchyma starts to aggregate. As reported previously (Betchaku, '67), the neoblasts have a peculiar mechanism of locomotion and their movement results in their mutual traction, or aggregation: a neoblast has a bipolarized form with distinct cytoplasmic processes, and possess strong isotypic adhesiveness. Fixed parenchyma cells have the most flexible cytoplasm and have ruffled membrane activities all around the edge of the cytoplasm, being different from the neoblasts which have active membrane only at the tip of the process (Betchaku, '67). Thus these fixed parenchyma cells can stretch extremely thin and long on the substrate, and can enter the smallest interstices in a cluster of neoblasts *in vitro*. The difference of the behavior of these two parenchyma cells is probably due to the difference in the distribution of active membrane on the surface of cells. We may regard neoblasts as accumulative cells and fixed parenchyma cells as interstitial cells. This behavior of fixed parenchyma cells is in good accord with their presumed function of the transport of metabolites in planarian parenchyma (Pederson, '61).

The affinity or the adhesiveness between cell types is of interest. Each of these tissues demonstrates not only difference in behavioral pattern, but also remarkably low heterotypic affinity or adhesiveness. This means that the more the tissues move, the more they tend to segregate themselves. In other words, the smaller the size an isolated body fragment is, the more complete is the tissue segregation. The first morphological behavior of tissues in an isolated body fragment is the detachment of tissues from their adjacent tissues, or the separation of the three layers (fig. 7). The low heterotypic tissue affinity seems to exaggerate their isotypic affinity. After two days *in vitro*, each tissue in the isolated body fragment segregates itself without mixing with other tissues.

An important cellular segregation occurs in the parenchyma, namely between fixed parenchyma cells and neoblasts. In the exposed area of partially or totally exposed parenchymal condensate of 22.5 hours and older fragments, the two cell types segregate with neoblasts in the periphery and fixed parenchyma cells arranged internally. In the area tightly covered with either epidermis or gastrodermis, the segregation does not occur. This observation suggests that the segregation between fixed parenchyma cells and neoblasts may be a reversible phenomenon operated by the difference of these two cell types in locomotion, in susceptibility, and in affinity. If they are exposed to a "harsh" condition such as the plain saline solution employed in the present experiment, the fixed parenchyma cells may either lyse or retreat deeper into the interior of the parenchymal mass while neoblasts remain at their original site and aggregate. The neoblasts remain at their site either because they may be more resistant to the environmental changes or because they may lack the free locomotive properties, and most likely because of both reasons. When the exposed parenchyma becomes covered with the spreading epidermis later in the process of regeneration or wound healing, fixed parenchyma cells may return to the periphery between neoblasts, and the distribution of these two cell types in the parenchyma may return to normal. This must occur not because the segregated neoblasts disperse themselves positively and spread evenly in the parenchyma, but very likely because fixed parenchyma cells forcibly enter between neoblasts to disperse them evenly as occurs *in vitro* (Betchaku, '67). To regain a sufficient population of fixed parenchyma cells later in the desegregation stage during the growth of a blastema, either or both proliferation of fixed parenchyma cells and differentiation from other type(s) of cells into fixed parenchyma cells are thought to occur.

Steinberg ('62, '64) proposed so far the most successful hypothesis of differential cellular adhesiveness for the sorting-out mechanism within a composite cell population employing a thermodynamic principle that the free energy of any system will tend toward a minimum. In the case of a binary composite aggregate of artificially dissociated-and-reaggregated cells, the more cohesive cell type aggregates internally while the less cohesive cell type aggregates externally. He assumes that the cells move randomly within the composite aggregate, and that the strength of the heterotypic adhesion is intermediate between those of the two isotypic associations. He describes a similar topography of segregation from conjugated undissociated tissue fragments, suggesting, here too, that the random mobility of cells and differential adhesiveness might be at work.

Recently, Roth and Weston ('67) circulated unlabelled isotypic aggregates or pellets, of liver or neural retina cells from seven day chick embryos in a homogeneous suspension of tritiated thymidine labelled liver or neural retina cells. When these two types are mixed into a composite aggregate, according to the authors, liver cells aggregate internally with respect to neural retina cells. According to Steinberg's theory, then, isotypic adhesion of liver cells should be greater than that of neural retina cells and that the heterotypic adhesion between these two types should be between that of the two isotypic adhesions. The adhesive strength was measured by the number of the labelled cells that adhered to the unlabelled cell pellets. The result was, however, not in agreement with the prediction by Steinberg's theory since isotypic association was found more stable than heterotypic association between these cell types. In the case of non-dissociated tissue fragment-fragment conjugation, Weston and Abercrombie ('67) reported, again in contradiction to Steinberg's assumption, that the intermingling of cells at the border between tritiated thymidine labelled and unlabelled fragments was negligible in both heteronomic and homonomic fusions between chick embryonic heart and liver fragments.

When the illustrations of both Roth and Weston ('67) and Weston and Abercrombie ('67) are compared, however, one can see an interesting difference between the topographies of labelled and unlabelled cells at the border of isotypic associations: the labelled cells from suspension which have adhered to the isotypic pellets are apparently penetrating to the subsurface region instead of making a uniform surface layer, whereas the border line of the fused isotypic fragments is very sharply defined. While this tendency of the attached cells to submerge into the isotypic aggregate projects a shadow of inadequacy over the method of Roth and Weston in measuring the rate of cellular adhesion (as first questioned by Mr. Albert Harris in our laboratory), the topographic difference of these two border lines indicates that the cellular susceptibility must be involved in making such a difference in cellular behavior. In the case of the fragment-fragment conjugation, the cells at the border line of fusion face no preferential gradient in environment

around them, whereas the cells adhering, from suspension, to the isotypic pellets may prefer the subsurface area of the isotypic cell aggregate to the culture medium.

The differential adhesiveness theory of Steinberg may explain the cellular segregation within planarian parenchyma in that the cells exposed to the culture medium change their adhesive properties in such a manner that the isotypic adhesiveness of fixed parenchyma cells is greater than that of neoblasts, and therefore neoblasts aggregate in the periphery of the fragment. The actual observations, however, did not agree with this explanation. If such drastic changes in their cellular adhesiveness ever occurs because of the exposure to the culture medium, similar changes should occur in the cells *in vitro* in the same culture medium. *In vitro* (Betchaku, '67), the fixed parenchyma cells do not aggregate and remain as individual active amoeboid cells which are attracted to the aggregating neoblasts rather than to their own kind, indicating that the isotypic adhesiveness of fixed parenchyma cells may be weaker than their heterotypic adhesiveness to neoblasts, and that they are interstitial cells in nature. Since neoblasts demonstrated their positive isotypic affinity, it can safely be described that, *in vitro*, the isotypic adhesion of neoblasts is greater than that between fixed parenchyma cells. It is apparent that the differences of cellular elements in their susceptibility and mobility, not only in their adhesiveness, are involved in the process of the cellular segregation and desegregation within planarian parenchyma.

During the course of tissue segregation, each tissue and cell type kept its identity quite clear. There was no evidence or even suspicion of dedifferentiation, or of differentiation of cells of any kind within the period of tissue segregation. Each cell type isolated from the cultured fragments of different ages, from 0 hour to 50.5 hours, did not show any difference in behavioral pattern, morphology, stainability *in vitro* (Betchaku, '67). Because a regeneration blastema of fresh-water planarians, which appears as an accumulation of neoblasts at the site of the wound, is formed also in about two days (Lindh, '57; Pederson, '58; Lender and Gabriel, '60), it is most unlikely that neoblasts found in a regeneration blastema of this age include dedifferentiated cells. The differentiation would occur probably after the second or the third day. The regeneration blastema must first occur as a parenchymal aggregate with an increased population of neoblasts due to both reduced number of more susceptible fixed parenchyma cells and the proliferation of more resistant neoblasts.

If an initial accumulation of neoblasts at the wound is the result of both the condensation of the local parenchyma and the segregation of its cellular components, instead of by the positive migration of neoblasts from near and distant areas to the site of the wound, the cellular composition of the blastema should not be so homogenous as some authors indicate (Rose and Shostak, '68). Some cells should lyse *in situ* to become motionless debris, and they should, with some of the less mobile unicellular inclusions of the parenchymal composite, be included in the condensing parenchyma or the blastema. My observation (unpublished) showed that between the accumulated neoblasts of an early regeneration blastema (1 to 2 day old) there were always cells other than neoblasts and their debris. There was not a single stage during the course of regeneration when the blastema was composed entirely of neoblasts. On the third day after the decapitation (at 17–18° C), the cells from the two cut stems of the longitudinal nerve cords were observed to have completed a definite juncture in the anterior blastema, suggesting active migration of various tissues into the blastema.

The cellular heterogeneity of three to four day old blastemata has been reported by electron microscopical investigations (Le Moigne et al., '65; Hay, '66). Sauzin ('67a,b) also observed differentiating nerve cells and muscles in a three day old blastema. Sengel ('60) observed that an isolated anterior blastema from a three day old regenerating planaria had differentiated into cephalic structures whereas a posterior blastema had not. She argued that the fate of a blastema, or more accurately, the fate of the neoblasts in the blastema is determined earlier. Such apparent tissue determination, however, may be explained by contamination of a blastema with cells of old tissues, nervous tissue in particular, which either remain in the parenchymal aggregate (blastema) or grow into the blastema. The fact that the formation of planarian regeneration blastema is not seriously impeded by low temperature whereas cytodifferentiation or regeneration is retarded (Brøndsted, '61) seems to favor the view that cytodifferentiation is not involved in the formation of a blastema within the first two days. The concept of the local origin of neoblasts in an early regeneration blastema now seems more realistic than the concept of positive and extensive migration of neoblasts, near and distant, to the wound. The growth of a blastema as well as its histogenesis, however, very likely involves the outgrowth of old tissues into this aggregate of neoblasts, as recently indicated by the study of thymidine kinase activity during regeneration (Coward, '69). The role of neoblasts in the histogenesis of a blastema, then, is to be left for further investigation.

Although no radioautographical investigation has been made, the mitotic figures were observed only among neoblasts in the cultured body fragments with increasing frequency toward the end of the second day. Some of the highly differentiated cells such as muscles, nervous tissue, and various gland cells exposed to the medium, (saline solution) showed signs of necrosis. They remain, however, identifiable throughout the course of tissue segregation. Planarian cell types of the adult organism may, therefore, be considered quite stable in nature and, in addition, these highly differentiated cell types do not seem to gain mobility. If there are transverse and the antero-posterior gradients in the rate of regeneration (Child, '11; Brøndsted, '46, '52, '56), the less mobile tissues, such as muscles and nerves, may play some roles in maintaining them. Even among the most mobile tissues, however, either in normal regeneration or in the intact body, there may not be much migration of cells, because of the extremely low heterotypic adhesion. This low rate of cellular mobility is probably one of the causes of the high frequency of heteromorphosis in planarian regeneration. The only exception to this cellular immobility *in vivo* may be the fixed parenchyma cells which may, as suggested by the previous report (Betchaku, '67), move freely in the parenchyma.

The amount of neoblasts aggregated in the fragments of approximately one fifth of the minimum size of a body fragment capable of regeneration shows that a sufficient amount of neoblasts for a blastema are present in each fragment (figs. 22–27). The physical separation between tissues undergoing segregating movement eliminates the possibility that dedifferentiated cells from other cell types may contribute to the population of neoblasts. This indicates that each minute portion of the planarian body has an independent population of neoblasts which can accumulate into a blastema without requiring additional neoblasts from distant areas. The mitotic activity of neoblasts, observed to increase toward the end of the second day (figs. 28–29), also indicates that the number of neoblasts may be increased by their proliferation. The difference between the behavior of neoblasts both *in vitro* (Betchaku, '67) and in the isolated body fragment and the behavior of neoblasts in the local wounded region of a whole animal may be small. When the animal is decapitated or injured in fresh-water, the surrounding area of the wound is exposed to a highly hypotonic condition with some possible buffering action from the exudate of the wound, which may be similar to the slightly hypotonic saline solution used for the culture. The similarity in morphology between singly isolated neoblasts, neoblasts in the cultured body fragment, and neoblasts *in situ* of an intact parenchyma also suggests similarity in behavior.

Inability of the small body fragments, used in the present experiment, to regenerate normally seems due to the failure of the aggregates of three tissue layers to make a proper stratification within the fragments. The larger fragments which regenerate normally will retain their gastrodermis internally and be covered with highly cohesive epidermis. Their parenchyma will be packed tightly between the two epithelial layers. These include the

one sixteenth cross-pieces of Child ('11) and their lateral half-pieces of Morgan ('01) which have more than five times as much volume as the body fragments used here.

To keep the accumulative parenchyma in the proper stratification, an external pressure applied on it from both sides seems necessary. Without an external force, planarian tissues, having very little heterotypic affinity, even seem to repel each other. In the larger regeneratable fragments, the expansion of gastrodermis may press the accumulative parenchymal composite against the strongly cohesive epidermal layer which covers the entire surface, and then, the mobile parenchymal cells may be able to fill every space between tissues and organs, creating the condition under which the transfer of metabolites between cells and the cytodifferentiation may be achieved. If a fragment has insufficient amount of epidermis to cover its surface, the gastrodermis may push its way out of the fragment, and the centrifugal pressure of the gastrodermis will be lost. The parenchymal mass concealed within an epidermal sheet alone (fig. 33), as is the case in an inverted fragment (figs. 9, 36), does not fill the space between tissues in the fragment.

ACKNOWLEDGMENTS

The author is indebted to Drs. B. V. Hall and F. J. Kruidenier of University of Illinois, Urbana, Illinois, for their kind courtesy of allowing him to use the facilities of their laboratories. The main text was prepared at the laboratory of Dr. J. P. Trinkaus of Yale University aided by a grant from Eli Lilly and Company, Indianapolis, Indiana.

LITERATURE CITED

Abercrombie, M., and E. J. Ambrose 1958 Interference microscope studies of cell contacts in tissue culture. Exptl. Cell Res., 15: 332–345.

Arnold, G. 1909 Intra-cellular and general digestive process in Planariae. Quart. Journ. Micro. Sci., 54: 207–220.

Bartsch, O. 1923 Die Histogenese der Planarienregenerate. Roux' Arch., 99: 187–221.

Betchaku, T. 1967 Isolation of planarian neoblasts and their behavior *in vitro* with some-aspects of the mechanism of the formation of regeneration blastema. J. Exp. Zool., 164: 407–434.

Brøndsted, H. V. 1939 Regeneration in planarians investigated with a new transplantation technique. K. Danske Vid. Selsk., Biol. Medd., 15: 1–39.

——— 1946 The existence of a static, potential, and graded regeneration field in planarians. K. Danske Vid. Selsk., Biol. Medd., 20: 1–30.

——— 1955 Planarian regeneration. Biol. Rev., 30: 65–126.

——— 1956 Experiments on the time-graded regeneration field in planarians with a discussion of its morphogenetic significance. K. Danske Vid. Selsk., Biol. Medd., 23: 1–39.

Brøndsted, A., and H. V. Brøndsted 1952 The time-graded regeneration field in planaria (Dugesia lugubris). Vidensk. Medd. Dansk. Natur. Foren. Kbh., 114: 443–447.

——— 1954 Size of fragment and rate of regeneration in planarians. J. Embryol. exp. Morph, 2: 49–54.

——— 1961 Influence of temperature on rate of regeneration in time-graded regeneration field in planarians. J. Embryol. exp. Morph., 9: 159–166.

Buchanan, J. W. 1933 Regeneration in Phagocata gracilis (Leidy). Phys. Zool., 6: 185–204.

Curtis, W. C. 1902 The life history, the normal fission and the reproductive organs of Planaria maculata. Proc. Boston Soc. Nat. Hist., 30: 515–559.

Child, C. M. 1911 Studies on the dynamics of morphogenesis and inheritance in experimental reproduction. I. The axial gradient in Planaria dorotocephala as a limiting factor in regulation. J. Exp. Zool., 10: 265–320.

Coward, S. J., J. H. Taylor and F. M. Hirsh 1969 Thymidine kinase activity during the regeneration of the planarian Dugesia dorotocephala. AM. Zoologist, 9: 611 (Abstract).

Dubois, Fr. 1949 Contribution à l'étude de la migration des cellules de régénération chez les Planaires dulcicoles. Bull. Biol. Fr. Belg., 83: 213–283.

Fedecka-Bruner, B. 1965 Régénération des testicules des planaires apres destruction par les rayons x. In: Regeneration in animals and related problems. V. Kiortsis and H. A. L. Tranpusch, eds. North-Holland Publ. Co., Amsterdam, 1965, p. 185–192.

Goetsch, W. 1921 Regeneration und Transplantation bei Planarien. Arch. f. Entw'mech., 49: 359–382.

Graff, L. von 1912–1917 In: Bronn's Klassen und Ordnungen des Tier-Reichs. Bd. 4, Ic: Turbellaria. II. Tricladida. C. F. Winter'sche Verlag, Leipsig.

Hay, E. D. 1966 Regeneration. Holt, Rinehart and Winston, Inc., New York, 1966.

Hyman, L. H. 1951 The invertebrates. Vol. 2. McGraw-Hill, New York, 1951.

Ijima, I. 1884 Untersuchunge über den Bau und die Entwicklungsgeschichte der Susswasser Dendrocoelen. Ztsch. Wiss. Zool., 40: 359–464.

Keller, J. 1894 Die ungeschlechtliche Fortpflanzung der Susswasser-Turbellarien. Jen. Zeit. Naurv. 28: 370–407.

Lash, J. W. 1955 Studies on wound closure in urodeles. J. Exp. Zool., 128: 13–28.

——— 1956 Experiments on epithelial migration during the closing of wounds in urodeles. J. Exp. Zool., 131: 239–256.

Lender, Th. 1952 Le rôle inducteur de cerveau dans la régénération des yeux d'une planaire d'eau douce. Bull. Biol. Fr. Belg., 86: 140–215.

Lender, Th., and A. Gabriel 1960 Etude histochimique des néoblasts de Dugesia lugubris (Turbellarie Triclade) avant et pendant la régénération. Bull. Soc. Zool. Fr., 85: 100–110.

Lewis, F. T. 1926 An objective demonstration of the shape of cells in masses. Science, 63: 607–609.

Li, Y. 1928 Regulative Erscheinungen bei der Planarienregeneration unter anomalen Bedingungen. Roux' Arch., 114: 226–271.

Lindh, N. O. 1957 Histological aspects on regeneration in Euplanaria polychroa. Ark. f. Zool., 11: 89–103.

Mattiesen, E. 1904 Ein Beitrag zur Embryologie der Süsswasserdendrocölen. Ztsch. Wiss. Zool., 77: 274–361.

Metschnikoff, E. 1883 Die Embryologie von Planaria polychroa. Ztschr. Wiss. Zool., 38: 331–354.

Minot, Ch. S. 1877 Studien an Turbeilarien. Beitrage zur Kenntniss der Plathelminthen. Arbeiten Zool-Zoot. Inst. Würburg, Bd. 3: 405–471.

Moigne, A. Le, M. J. Sauzin, T. Lender et R. Delanault 1965 Quelques aspects des ultrastructures du blasteme de régénération et des tissus voisins chez Dugesia gonocephala (Turbellaire, Triclade). C. R. Soc. Biol., 159: 530–534.

Morgan, T. H. 1901 Growth and regeneration in Planaria lugubris. Arch. EntwMech. Org., 13: 179–212.

Okada, Yô K., and T. Kido 1943 Further experiments on transplantation in planaria. J. Fac. Sci. Imper. Univ. Tokyo, Sec. IV, Zool., 6: 1–23.

Okada, Yô K., and H. Sugino 1934 Transplantation experiments in Planaria gonocephala. Proc. Imper. Acad. Jap., 10: 107–110.

——— 1937 Transplantation experiments in Planaria gonocephala Duges. Japan. J. Zool., 7: 373–439.

Pedersen, K. J. 1958 Morphogenetic activities during planarian regeneration as influenced by triethylene melamine. J. Embryol. exp. Morph., 6: 308–334.

——— 1959 Cytological studies on the planarian neoblast. Zeitsch. f. Zellforsch., 50: 799–817.

——— 1961 Studies on the nature of planarian connective tissue. Zeitsch. f. Zellforsch., 53: 569–608.

Polezhaev, L. W., and N. I. Ermakowa 1960 Restoration of regeneration capacity of axolotls which was suppressed by x-ray. (in Russian) Doklady Acad. Sci., USSR, 131 (1): 209–214.

Polezhaev, L. W., N. A. Teplic and N. I. Ermakowa 1961 Restoration of regenerative capacity of extremities of axolotl supressed by exposure to x-rays, by proteins, nucleic acids and lyophilized tissues. (in Russian) Doklady Acad. Sci., USSR, *138:* (2): 477–480.

Prenant, M. 1922 Recherches sur le parenchyme des Plathelminthes. Essais d'histologie comparée. Arch. Morph. gen. exp., *5:* 1–72.

Rose, C., and S. Shostak 1968 The transformation of gastrodermal cells to neoblasts in regenerating *Phagocata gracilis* (Leidy). Exp. Cell. Res., *50:* 553–561.

Roth, S. A., and J. A. Weston 1967 The measurement of intercellular adhesion. Proc. Nat. Acad. Sci., *58:* 974–980.

Sauzin, M. J. 1967a Etude ultrastructurale de la différenciation du néoblaste au cours de la régénération de la Planaire *Dugesia gonocephala*. I. Différenciation en cellule nerveuse. Bull. Soc. Zool. Fr., *92:* 313–318.

———— 1967b Etude ultrastructurale de la différenciation au cours de la régénération de la Planaire *Dugesia gonocephala*. II. Différenciation musculaire. Bull. Soc. Zool. Fr., *92:* 613–616.

Sengel, C. 1960 Culture *in vitro* de blastèmes de régénération de planaires. J. Embryol. exp. Morph., *8:* 468–476.

Sivickis, P. B. 1932 A quantitative study of regeneration along the main axis of the triclad body. Arch. Zool. ital., *16:* 430–449.

Steinberg, M. S. 1962 Mechanism of tissue reconstruction of dissociated cells. II. Time-course of events. Science, *137:* 762–763.

———— 1964 The problem of adhesive selectivity in cellular interactions. In: Cellular membranes in development. Michael Locke, ed. Academic Press, N. Y., pp. 321–364.

Stephane-Dubois, Fr. 1961 Les cellules de régénération chez la planaire *Dendrocoelum lacteum*. Bull. Soc. Zool. Fr., *86:* 172–185.

Sugino, H. 1936 Inhibition of anterior regeneration by the coalescence of wound in *Planaria gonocephala*. (in Japanese text with illustrations) Kagaku, *6:* 138–139.

Thompson, D. 1961 On Growth and Form. J. T. Bonner, ed. Cambridge Press, Cambridge, England, pp. 119–125.

Vaughan, R. B., and J. P. Trinkaus 1966 Movements of epithelia cell sheets *in vitro*. J. Cell Sci., *1:* 407–413.

Weston, J. A., and M. Abercrombie 1967 Cell mobility in fused homo- and heteronomic tissue fragments. J. Exp. Zool., *164:* 317–324.

Weiss, P. 1959 The biological foundations of wound repair. In: The Harvey Lectures. Series 55, 1959-60. Academic Press, New York and London, 1961, pp. 13–42.

Willier, B. H., L. H. Hyman and S. A. Rifenburgh 1925 A histochemical study of intercellular digestion in triclad flatworms. J. Morph. Physiol., *40:* 299–340.

Wolff, Et., and Fr. Dubois 1948 Sur la migration des cellules de régénération chez les planaires. Rev. Suiss Zool., *55:* 218–227.

Wolff, Et. 1962 Recent researches on the regeneration of planaria. In: Regeneration. The Ronald Press Co., New York, pp. 53–84.

PLATES

PLATE 1

EXPLANATION OF FIGURES

2   A mid section of one of the body fragments fixed within five seconds after being cut from the prepharyngeal region of the animal in saline solution. Each fragment contains almost equal varieties of cells and tissues. Fixed in a formalin-alcohol-acetic acid mixture, paraffin sectioned, and stained with Delafield's haematoxylin and eosin. vep, ventral epidermis; dep, dorsal epidermis; gt. gastrodermis; pa, parenchyma; nv, nerves; mu, muscles. × 115.

3   A mid section of another sample of a similar body fragment fixed within five seconds after being cut out in saline solution. × 115.

4   Gastrodermal cells from a 22.5 hour cultured fragment. The highly vacuolated and enlarged migrated gastrodermal cells illustrated in the following figures (figs. 6–32) can be easily dissociated from the gastrodermal sheet by being pressed to the bottom of the culture dish, where they resume normal appearance. Fixed in Zenker's fixative *in toto* on a cover glass previously placed on the bottom of the culture dish. Arrows indicate nuclei. (Reproduced from a color transparency of a preparation for PAS reaction; the nuclei are stained with Mayer's haemalaun). × 560.

5   Gastrodermal cells, dissociated by crude trypsin from a fresh body fragment, were pressed under a cover glass and photographed. Being flattened, the cells were exaggerated in size. Varied number of food vacuoles with or without ingested material are one of the characteristic features of these phagocytic cells. Arrows indicate nuclei. Phase-contrast. × 460.

6   A ciliated epidermal cell from the ventral epidermis, dissociated by crude trypsin from a fresh body fragment, moves vigorously with its beating cilia. Live, in saline solution. Phase-contrast. × 460.

PLATE 1

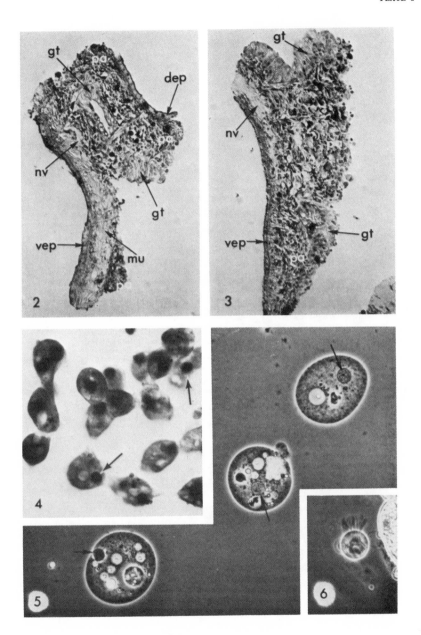

PLATE 2

EXPLANATION OF FIGURES

All specimens were cultured in saline solution, then fixed at intervals indicated in a formalin-alcohol-acetic acid mixture, paraffin sectioned, and stained with Delafield's haematoxylin and eosin. Abbreviations: gt, gastrodermis; pa, parenchyma; ep, epidermis; mu, muscles; nv nerves.

7 A mid section of a five hour cultured fragment. The gastrodermis has detached from the parenchyma and is stretching from the interior of the fragment. Such gastrodermal cells loose their polarity and columnar shape. The epidermis also detaches from the basement membrane. Arrows point to the stretching fronts of the gastrodermis. $\times$ 114.

8 A mid section of a 10.5 hour cultured fragment. The gastrodermis has stretched out completely creating a huge cavity between it and the underlying parenchyma. The large arrows point to the non-advancing and thickened frontal ends of the epidermis which has also detached from the basement membrane. The broken line indicates a portion of the stretched gastrodermis which was torn off during dehydration. $\times$ 114.

9 A mid section of a 22.5 hour cultured fragment in which total topographic inversion has occurred. Although inverted, the epidermis, unlike gastrodermis, retains its polarity. Note cavities between parenchyma and gastrodermis. Tissues and cells are found loosened in these cavities, one of which is marked by X. $\times$ 114.

10 A part of a cultured fragment 22.5 hours old showing an inverted epidermal aggregate and migrated gastrodermis. $\times$ 96.

11 An enlargement of figure 10. Although the epidermis has been detached from the underlying tissues and clumps together, the relative stratification of tissues has not been altered. $\times$ 280. The total view of the fragment is shown in figure 36.

PLATE 2

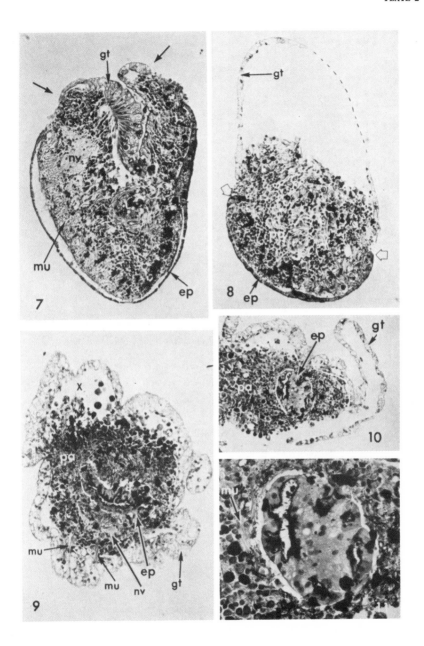

PLATE 3

EXPLANATION OF FIGURES

Technique and abbreviations as in plate 2.

12–18   The adhesion between gastrodermis and epidermis is shown in
series of 5 μ thick sections of a 22.5 hour cultured fragment whose
entire mid section is shown in figure 12. Highly vacuolated gastrodermis seems to stick well to the cilia of the epidermis, but the
lateral sides of the epidermis are undoubtedly in tighter contact
with the gastrodermis. Only the difference of cytoplasmic density
distinguishes these two cell types in cohesion in these preparations. Figure 12: $\times$ 113. Figures 13–18: $\times$ 257.

19   The affinity between gastrodermis and epidermis, either ciliated
or non-ciliated, apparently higher than that of between gastrodermis and parenchyma. The migrated gastrodermis of this 27.5
hour cultured fragment adheres tightly to the epidermis while
on the right side stretching over the parenchyma without attach-
ing to it (indicated by arrows). $\times$ 72.

20   A mid section of another 27.5 hour cultured fragment. The epidermis aggregates on the surface of the fragment. The gastrodermis stretched over almost the entire fragment. The differential
adhesion of gastrodermis with epidermis and with parenchyma is
shown in the thinly stretched gastrodermis over the parenchyma
(arrows) and the thick and non-stretched gastrodermis over the
epidermis. $\times$ 113.

21   A mid section of a 32 hour cultured fragment. The epidermis, instead of stretching over the fragment, aggregates into folds. Very
few parenchyma cells are seen under the clumping epidermis
indicating that the parenchyma is also aggregating. $\times$ 114.

PLATE 3

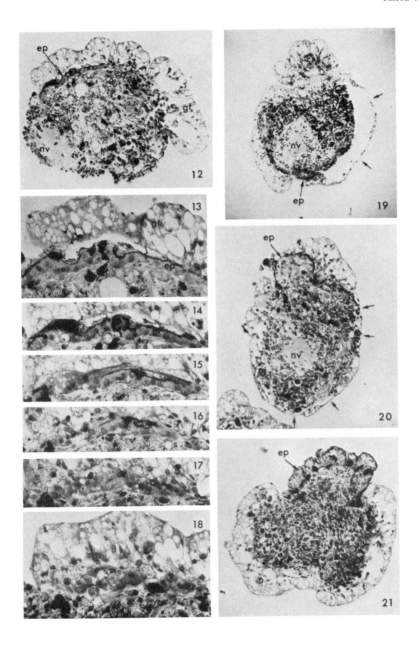

PLATE 4

EXPLANATION OF FIGURES

Technique as in plate 2. Abbreviations: ep, epidermis; gt, gastrodermis; nv, nervous tissue; mus, muscles.

22 A mid section of a 27.5 hour cultured fragment. Segregation of tissues is in progress. Under the clumping epidermis, only non-mobile muscles and nervous plexi remain with vacant space around them. × 116.

23–25 The separation of epidermis and parenchyma is illustrated in three sections from the same fragment of 32 hour old. Figure 23 — median plane section. × 143.

26 In this 27.5 hour cultured fragment, the segregation of tissues is reaching completion. In the absence of a protective layer of either epidermis or gastrodermis, the segregation of neoblasts is maximal due to the difference in susceptibility of fixed parenchyma cells and of neoblasts to the culture medium (plain salt solution). Large arrows show aggregating neoblasts. × 214.

27 Another section of the fragment of figure 26. A typical orientation of aggregating neoblasts with processes is seen at upper right (a large arrow). A long cytoplasmic process of one of the neoblasts extending outward is indicated by a small black arrow. × 214.

PLATE 4

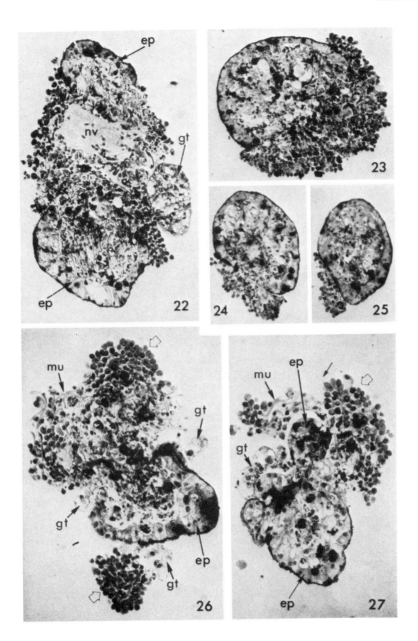

PLATE 5

EXPLANATION OF FIGURES

Techniques as in plate 2. Abbreviations: ep, epidermis; gt, gastrodermis; pa, parenchyma; mu, muscles; nb, neoblasts; nv, nerves.

28–29 Mitotic figures (arrows) are observed only among the neoblasts during the first three days in culture, at especially high frequency in the third day. Figure 28 ($\times$ 436) shows a portion of the parenchyma of a 50.5 hour cultured fragment. Figure 29 ($\times$ 125 shows the entire section of the fragment.

30–31 An epidermal aggregate on the surface of a non-inverted fragment. Migrating outward from a small body fragment, the gastrodermis usually positions itself on the surface of the fragment, and keeps the epidermal aggregate from departing from the main mass of the body fragment by adhering to it. $\times$ 213.

32 In this 22.5 hour cultured fragment, the segregation of tissues occurred rapidly. A portion of the gastrodermis formed an empty gastrodermal tube. The epidermal sphere contains only non-mobile elements such as muscles and nerve plexi which innervate the epidermis. The mobile cells and tissues have segregated isotypically. $\times$ 128.

33 A mid section of a 27.5 hour cultured fragment encircled by the epidermis. No segregation of tissues occurs inside the epidermal coverage. The gastrodermis seems to have migrated from the sphere. Cells and tissues are loosely distributed without tight contact with the epidermis. $\times$ 113.

62

PLATE 5

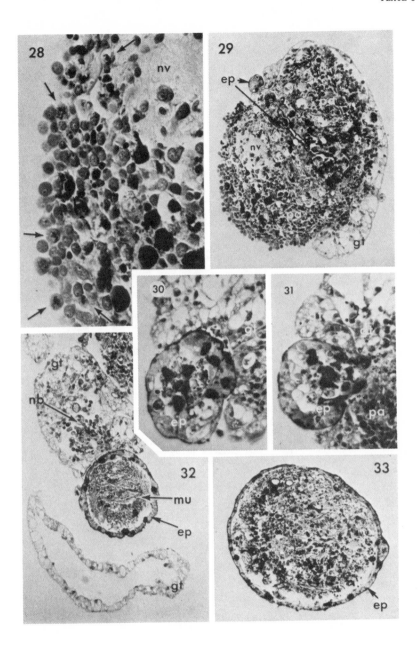

PLATE 6

EXPLANATION OF FIGURES

Technique as in plate 2. Abbreviations: vep, ventral epidermis; dep, dorsal epidermis; pa, parenchyma; gt, gastrodermis; nv, nervous tissue.

34 A mid section of a 50.5 hour cultured fragment. The gastrodermis has migrated, and the epidermis encloses the rest of tissues. The detachment of both ventral and dorsal epidermis from the underlying tissues is apparent (the fate of the basal membrane is unknown). The flattening of epidermal cells is more prominent in the dorsal epidermis, and an accumulation of epidermal cells (large arrow) at the margin is observed. A small arrow points to the original lateral edge of the epidermis. × 120.

35 Enlargement of the marginal end of the dorsal epidermis of figure 34. × 198.

36 Two conjugated fragments cultured for 22.5 hours. The spreading of epidermis is more or less disturbed. The lower fragment has been split open in the middle of the gastrovascular cavity. The structure of the dorsal half of the lower fragment has been inverted. The contact area (approximately trimmed by the black rectangle) is enlarged in figures 27–38, although in different sections. × 104.

37 The contact area of the two fragments through parenchyma. Neoblasts have accumulated in the area, and are responsible for parenchymal conjugation. × 167.

38 The contact area of the two fragments through nervous tissue. In this section, the nervous tissue seems to be the most tightly attached tissue between two fragments. × 167.

64

PLATE 6

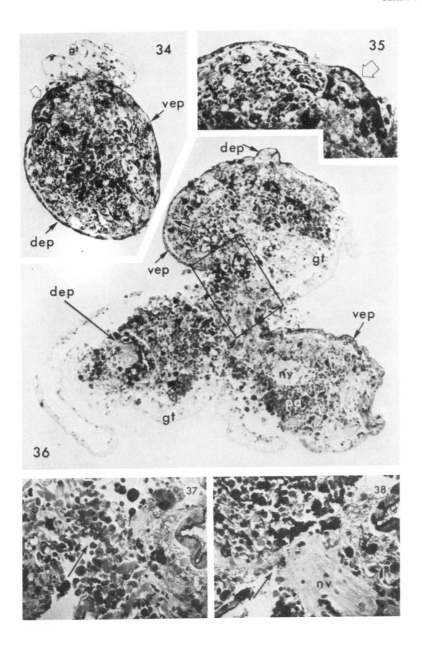

THE EFFECT OF AC FIELD FREQUENCY ON THE
REGENERATION AXIS OF *Dugesia tigrina*[1]

Gordon Marsh

Introduction

Alternating current at 60 cycles/second ($=60$ Hz) was found to produce head behavior in the morphological tail, incipient structural bipolarity with subsequent regression of the posterior "head," or permanent bipolarity, at field strengths of from 310 to 418.4 rms mv./mm. (Marsh, '68). Polarity was not reversed at any field strength. The field effects were non-polar, but occurred only when the regenerant's antero-posterior axis and the field axis coincided. No rectification or other distortion of wave form was produced by passage of the current through the regenerants.

Unlike the DC field, in which regenerant orientation with respect to the electrical poles is critical for polarity control (Marsh and Beams, '52), the AC field operates only to reduce or obliterate the normal polarity. In the DC field of a given strength the head-determining power of the cathode is greater than the tail-determining power of the anode. This fact, coupled with the necessity for parallelism of field and regenerant axes at 60 Hz, suggests the possibility that AC field action

[1] Supported by a National Science Foundation Grant (GB-4739).

is an overbalancing of the normal polarity by the appropriate half wave. An alternative mechanism might be the reduction of the polar difference by stimulation of respiration (Flickinger and Blount, '57) with accompanying alterations in the metabolic relations of polarity. In either case, and, indeed, from the circumstance of a positive influence of a symmetrical AC field on axial polarity, it becomes of interest to explore the relation of frequency to morphogenetic control.

MATERIALS AND METHODS

Three or more days after feeding, individuals of the planarian, *Dugesia tigrina* were sectioned transversely to produce 4 to 8 pieces with two cut surfaces, allowed to heal 15 to 29 hours, and mounted in 1.9% ionagar at 46 to 51° C. They were set in a rectangular chamber at 23 ± 1° C in aerated, flowing medium of 5000 ohm-cm. specific resistance, with axes parallel to the field direction and anterior ends oppositely directed in about equal numbers. At intervals they were reoriented, save for 15 regenerants which turned at right angles to the field direction and were allowed to remain. AC current of known frequency, and field strength of from 314.5 to 457 rms mv./mm. (the limits found effective at 60 Hz (Marsh, '68)), was applied for 4.1 to 5.8 days, at which time regenerants were freed from the agar and examined. (For additional details of method see Marsh, '68).

A Hewlett-Packard 233A audio oscillator supplied AC current at 6 frequencies from 5000 to 500,000 Hz. The frequencies were arbitrarily chosen; the high values represent a search for a frequency producing no effect. The oscillator output was a pure sine wave at all frequencies employed, insofar as could be judged by display on a Tektronix 502 oscilloscope. The latter was also used to verify the accuracy of the oscillator dial setting. No rectification or distortion of wave form occurred at any frequency for current passing through the regenerants, as shown by comparison of that current (in a special chamber) with the oscillator output on the dual beams of the Tektronix 502 (Marsh, '68).

Current strength was measured on a Weston model 622 variable thermo milliammeter, except at 100 and 500 Khz, where it was calculated from the potential difference developed across a standard resistor and measured on a Dumont 304H oscilloscope. The manufacturer's rating limits the Weston 622 milliammeter to frequencies of 15

Khz and below, but it was found usable to 50 Khz with recalibration of the scale.

RESULTS

Table 1 presents the effects of different field strengths on the morphogenetic state of regenerants at the frequencies appearing in the first column. Here, and subsequently, the data at 60 Hz are from Marsh, '68. Blank spaces occur where no pieces were exposed. The morphogenetic effects are (1) permanent two-headed condition, (2) temporary visible head structures and head behavior in the morphological tail, and (3) temporary head behavior in the morphological tail (Marsh and Beams, '52). As may be seen in the last column, permanent bipolar individuals were produced at all frequencies. The numbers appear to diminish irregularly with increasing frequency, but when expressed as percentages of surviving individuals they rise irregularly to a maximum at 25 Khz, then fall irregularly at the higher frequencies. The motility of imbedded regenerants was high at all frequencies, making conclusions derived from comparisons of figures of either category questionable.

The distribution of numbers of bipolar individuals with field strength showed no consistent trend at any frequency, save possibly at 500 Khz where bipolars were obtained only at the lowest field strength interval. The differences in total numbers of bipolars for the six field strength intervals shown at the bottom of Table 1, virtually disappear when expressed as percentages of survivors. The mean field strength at which bipolars were produced for frequencies above 60 Hz was 365.9 rms mv./mm.

Of the 77 bipolar regenerants produced at the higher frequencies, 34 had been oriented to the left end of the chamber, 42 to the right, and one at right angles to the field direction. The latter was the single individual of 15 so oriented to show any effect of the AC field. 60 regenerants showed dominance of the head developing on the original anterior end, 6 displayed no dominance, and in 11 cases dominance was not noted. The cases of no dominance were twice as numerous in the lower half of the field strength range as in the upper half; those of original anterior dominance were twice as numerous in the upper half as in the lower.

The number of regressive bipolars shows no obvious relation to field strength in Table 1, although slightly more than half of the total were found at the higher field strength intervals, which is reflected in the mean value of 386.1 rms mv./mm. for the higher frequencies. Similarly, the distribution with frequency shows no systematic trend. The distribution of numbers of functional bipolars with field strength and with frequency is similar to the corresponding distributions of numbers of permanent bipolars. The mean field strength for production of functional bipolars was 367.5 rms mv./mm. (excluding 60 Hz).

There was no relation between any of the three morphogenetic effects of the AC field at and above 5 Khz and either healing time or exposure time. This result is to be expected, since a similar independence was found at 60 Hz, where both healing time and exposure time varied over much wider ranges.

Regenerant survival in the AC field at frequencies of 5 Khz and above was substantially greater than that obtaining at 60 Hz or in DC fields (Marsh and Beams, '52). In figure 1A percentage survival is plotted against log Hz; the crosses are the mean per cent survival at each frequency, dots are percentages at different field strengths. The descent to a minimum at 25 Khz is believed attributable to technical factors rather than to frequency. Below 10 Khz regenerant mounting was done by one individual; at that frequency approximately half the regenerants were mounted by a second individual who mounted all regenerants at the higher frequencies. The decline and recovery of percentage survival probably represents learning and mastery of the mounting technique. If this may be assumed, the survival curve rises from 17.9% at 60 Hz to become asymptotic to approximately the 90% ordinate. The mean survival for the six high frequencies was 71.8%. Curves of survival against log Hz for the different field strength intervals roughly duplicate the form of the mean survival curve.

The improved survival in the high frequency AC fields over that found in the 60 Hz and DC fields is further shown in figure 1B. Curve 2 of that figure is a plot of percentage survival at 60 Hz against field strength in, usually, 25 rms mv./mm. intervals. The decline of survival from a high value at low to a plateau at intermediate field strength and the subsequent fall to zero at high field strength is similar in form, although differing in field strength range, to that obtained with DC fields (Marsh and Beams, '52). On the other hand the average survival

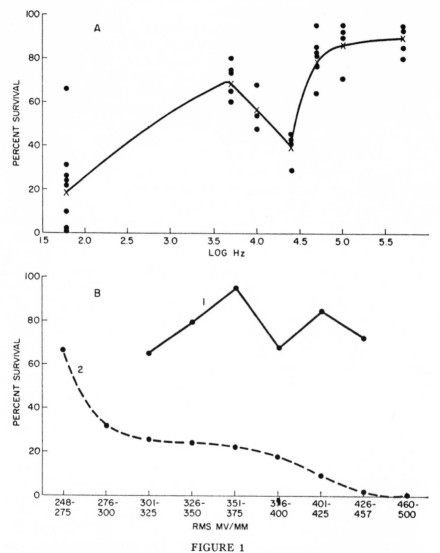

FIGURE 1

A. Percentage survival of regenerants vs. log frequency. Crosses: average values; dots: percentage survival at different field strengths.

B. Percentage survival vs. field strength in intervals of (usually) 25 rms mv./mm. Curve 1: 5-500 Khz. Curve 2: 60 Hz.

70

at 5 Khz and above, figure 1B, curve 1, shows only irregular variation about the 75% ordinate line when plotted against field strength interval, which may probably be interpreted to indicate that survival is virtually independent of field strength.

For the regenerants showing the three morphogenetic states in Table 1, the distribution with respect to body level of origin was of a pattern different from that found at 60 Hz or in DC fields. For each state the number of individuals was least for anterior pieces and greater for posterior, with the region at and just behind the posterior half of the pharynx contributing the greatest number. The distribution of surviving regenerants with respect to body position of origin was also differ-

TABLE 1

EFFECT OF FIELD STRENGTH AND FREQUENCY ON REGENERANT MORPHOGENETIC STATE

Hz	Number Mounted	Morphogenetic Effect*	Field Strength—rms mv./mm.						Total
			301-325	326-350	351-375	376-400	401-425	426-457	
60	2,265	(1)	2	10	5	8	4	0	29
		(2)	0	2	3	2	1	0	8
		(3)	2	9	9	11	1	0	32
5×10^{-3}	180	(1)	5	3		8	0	3	19
		(2)	0	0		2	1	0	3
		(3)	2	2		2	0	0	6
1×10^{-4}	181	(1)	4	1		3		4	12
		(2)	0	0		0		0	0
		(3)	2	0		0		0	2
2.5×10^{-4}	245	(1)	3	4		6		5	18
		(2)	2	0		1		0	3
		(3)	2	0		2		1	5
5×10^{-4}	328	(1)	1	3	0	1	3	1	9
		(2)	0	0	0	0	0	0	0
		(3)	3	8	1	2	4	3	21
1×10^{-5}	219	(1)	2	3	2	3	1	4	15
		(2)	0	1	1	1	0	4	7
		(3)	1	8	2	0	2	5	18
5×10^{-5}	217	(1)	4	0		0		0	4
		(2)	1	1		2		2	6
		(3)	3	0		1		2	6
Total		(1)	21	24	7	29	8	17	106
		(2)	3	4	4	8	2	6	27
		(3)	15	27	12	18	7	11	90

* (1) Permanent bipolars; (2) Regressive bipolars; (3) Functional bipolars.

71

ent from any previously encountered. With the center cut dividing the pharynx, the four major body regions were represented by nearly identical numbers of survivors.

No differences were found at any frequency (including 60 Hz), nor at any field strength, which were associated with regenerant size.

DISCUSSION

The production of axial morphogenetic alteration in regenerating *D. tigrina* at frequencies of 5 Khz and above is essentially similar to the effects at 60 Hz, except for extension of production to somewhat higher field strengths. Presumably this is an aspect of the differences in survival of exposed regenerants. A possible exception may exist at 500 Khz, where the number of regenerants with altered axial state was considerably less than at lower frequencies, as was also true when expressed as percentage of survivors. At 500 Khz the morphogenetic control exerted by the AC field may be declining, and at still higher frequencies it might be found ineffectual. If this were true, it would be expected that the data of Table 1 would have shown a clearer trend. There does exist the vague possibility that the rise in bipolar numbers, expressed as percentages of survivors, to a maximum at 25 Khz represents the reduction in mortality with increasing frequency, and that the decline to 500 Khz represents the diminution in effectiveness of the AC field. Because of the great motility of imbedded regenerants, which reduces the average axial field strength imposed, the number of individuals necessary to produce statistically significant values for the different frequencies appears forbiddingly large. Whatever the true relation to frequency, the mean field strengths at which permanent (365.9 mv./mm.) and functional bipolars (367.5 mv./mm.) were produced for the entire experimental population were the same as those at 60 Hz (364 and 363.6 mv/mm. respectively). The mean for regressive bipolars was 386.1 mv./mm. compared to 369.4 at 60 Hz.

The necessity for coincidence of the regenerant axis with field direction found at 60 Hz (Marsh, '68), appears supported at the higher frequencies, although the numbers exposed at right angles were small. A bipolar individual appearing at 5 Khz and 381.1 rms mv./mm. field strength was the single regenerant of 15 oriented at right angles to the field direction to show an effect. All 15 survived. As in the case of the similar individual at 60 Hz (Marsh, '68) there was no record of it moving from the right angle orientation. No regressive or functional bipolar regenerants appeared in this group; this was also true at 60 Hz. In view of the frequency of appearance of these forms in Table 1 and in DC fields (Marsh and Beams, '52; Dimmitt and Marsh, '52), it appears probable that these two permanent bipolars represent procedural or recording errors. It has long been suspected that regenerants which appear to maintain proper orientation in either DC or AC fields, but which regenerate normally, have probably spent a major fraction of the exposure time unobserved at angles to the field axis. It has been noted, particularly with DC fields, that regenerants which turn frequently tend to develop normally. It is possible that the two individuals discussed above turned into the field axis without being observed, and that this explains the bipolar condition, although in no instance has a regenerant beginning exposure at right angles to the field direction been observed so to turn.

The suggestion that the mechanism of action of the AC field is an overbalancing of the normal polarity by the appropriate half-wave, stems from the demonstration that in the DC field the head-determining power of the cathode is greater than the tail-determining power of the anode for the same field strength (Marsh and Beams, '52). In the AC field, irrespective of orientation, each regenerant with antero-posterior axis parallel to the field direction would experience a "cathodal" half-wave from posterior to anterior during each cycle, which would exert a greater head determination on the posterior than its anodal character in the opposing direction would exert as tail determination (head suppression), promoting development of the bipolar condition if the field strength were sufficient. This result would necessitate that the following half-wave would be ineffectual, or at least less effective, in producing the opposite result. The following half-wave would be cathodal for the anterior to posterior direction. In the DC field regeneration of individuals oriented anterior end to the cathode develop normally (Marsh and Beams, '52) with no evidence of effect of the applied current on any aspect of regeneration. While this is uncertain evidence of the ineffectiveness of antero-posterior cathodal current, it does suggest the possibility that a difference between the AC half-waves may obtain. A further necessity for the half-wave effect would be that head determination, once started at the posterior end, should continue, and that the anterior character bestowed upon the posterior end by the developing head should not reverse the relation and so neutralize the initial effect. Such a unidirectional developmental momentum would be in harmony with the all-or-none character of the effect of the DC field (Marsh and Beams, '52).

The half-wave mechanism is supported, although not compellingly, by the necessity for coincidence of the regenerant axis and the field direction, which would appear to make it improbable that the AC field action is of the character of an indifferent stimulus. It is also supported to a degree by the fact that the tendency of the AC field is to produce two-headed, not two-tailed, regenerants, so that whatever the nature of the field action its head-determining tendency exceeds its tail-determining tendency. The argument is somewhat weakened by the distribution of dominance with field strength. Dominance of the head regenerating on the original anterior end would be expected to be greatest at low and least at high field strength producing bipolar individuals, because of the persistent organization in the piece at isolation. This was approximately true for DC fields (Marsh and Beams, '52) over the field strength range from regressive bipolars to reversed axis regenerants, although within small field strength intervals discrepancies occurred. In the data presented in Table 1 the relation appears inverted; anterior head dominance was twice as frequent in the upper half of the field strength range as in the lower, while the cases of no dominance (head equality) were twice as frequent in the lower half as in the upper. The meaning of these differences is perhaps questionable, since the frequency of appearance of morphogenetic effect of the AC field was substantially independent of field strength. At 60 Hz, where the influence of field strength was more marked (Marsh, '68), dominance of the original anterior end was equally distributed over the field strength range; all cases of no dominance were found in the two lower field strength intervals.

The argument developed above is highly speculative, involving assumptions nearly or completely impossible to test. Perhaps the most telling evidence against the half-wave effect is the apparent absence of a relation of field effect to field strength, a relation which should resemble qualitatively that obtaining in DC fields if the "cathodal" half-wave is to be operative and the following half-wave ineffective. It was demonstrated for DC fields that the molecular mechanisms of control must lie in some combination of dipole orientation, valence change at critical energy levels, and establishment of a new electro-polarity (Dimmitt and Marsh, '52; Marsh, '62). The latter two modes of control necessitate connecting relations between the metabolic mechanisms of establishment and control of cellular electromotive force and the metabolic events of morphogenesis. There appears to be nothing in the data of morphogenetic effects of AC fields at any of the tested frequencies inconsistent with the view that the AC current produces such perturbations in reaction rates and states of metabolic constituents that polar differences are diminished. The effect of AC fields would then be analogous to the effects of metabolic inhibitors (Flickinger, '59; Flickinger and Coward, '62), which appear to promote bipolarity by reduction of the normal inherent polar difference between the ends of regenerants. This would appear more probable than the electrical establishment of a new polarity by the AC field.

Survival in the AC field increases with increasing frequency; at the lowest frequency, 60 Hz, survival has improved over that in DC fields. Strictly comparable data are not available for DC fields; in a 7095 ohm-cm. medium over a range of from 169.5 to 242 mv./mm. survival was 30.8 per cent, while survival at a slightly higher field strength at 60 Hz, figure 1, curve 2, was 66 per cent (Dimmitt and Marsh, '52). The maximum DC field strength at which a regenerant survived was 320 mv./mm.; this individual failed to develop symmetrically, and histolyzed within 4 days of removal from the field. The corresponding maxima for 60 Hz and 100 Khz were 437 and 457 rms mv./mm. respectively. Survival was zero at 60 Hz for field strengths to 500 rms mv./mm.; field strengths above 457 rms mv./mm. were not employed at the higher frequencies. From figure 1A it would appear probable that over this range of field strengths survival at still higher frequencies would approach 100 per cent, or equality to that of controls exposed to no field.

The most unexpected difference in survival tendency at high frequencies over that in DC fields and at 60 Hz, was in the distribution of survivors with respect to body level of origin. The higher mortality of pieces at and adjacent to the original pharynx had been thought attributable to the presence of the pharynx or of the pharyngeal cavity, which made healing more difficult and rupture due to movement more frequent. This appears less probable from the data at high frequencies, where survivors were substantially identical in numbers from each body level. The tendency to movement, based on comparisons of deviations from initial orientation in the chamber or of behavior in the special chamber used to test for wave form distortion, was conspicuously greater in the AC fields. The movement phenomena at 60 Hz were similar to those at higher frequencies, but the body level distribution of survivors resembled that of DC fields. If the survival differences are attributable to movement, the necessity for greater rotation to avoid the effects of DC fields may be sufficient to account for the difference between DC and AC survival with the mid pieces showing the greatest improvement. The 60 Hz stimulus to movement would then be greater than at higher frequencies, though less than that by DC fields.

References

1. DIMMIT, J., & MARSH, G. 1952. Electrical control of morphogenesis in regenerating *Dugesia tigrina*. II. Potential gradient vs. current density as control factors. *J. Cell and Comp. Physiol.*, **40**, 11-24.

2. FLICKINGER, R. A. 1959. A gradient of protein synthesis in planaria and reversal of axial polarity of regenerates. *Growth*, **23**, 251-271.

3. ———, & BLOUNT, R. W. 1957. The relation of natural and imposed electrical potentials and respiratory gradients to morphogenesis. *J. Cell and Comp. Physiol.*, **50**, 403-422.

4. ———, & COWARD, S. J. 1962. The induction of cephalic differentiation in regenerating *Dugesia dorotocephala* in the presence of normal head and in unwounded tails. *Develop. Biol.*, **5**, 179-204.

5. MARSH, G. 1962. Volume resistivity of regenerating *Dugesia dorotocephala* and the electrical work of polarity control. *J. Cell and Comp. Physiol.*, **59**, 273-280.

6. ———. 1968. The effect of sixty-cycle AC current on the regeneration axis of *Dugesia*. *Jour. Exp. Zool.*, **169**, 65-69.

7. ———, & BEAMS, H. W. 1952. Electrical control of morphogenesis in regenerating *Dugesia tigrina*. I. Relation of axial polarity to field strength. *J. Cell and Comp. Physiol.*, **39**, 191-213.

Modification of Regeneration in *Dugesia tigrina* by Actinomycin D

KRYSTYNA D. ANSEVIN AND MARILYN A. WIMBERLY

Members of the genus *Dugesia* can reconstitute a complete organism from a small fragment of their body. Several investigators (Santos, '31; Miller, '38; Okada and Kido, '43; Lender, '52) have observed that processes which occur in planarian regeneration apparently are similar to inductions and inhibitions in embryonic development. However, other than some evidence that in this group of animals the nervous system may be the source of inductors (Lender and Gripon, '62; Kido, '52; Teshirogi and Jin, '64) as well as of inhibitors (Török and Törö, '62), nothing is known about the mechanism of either phenomenon. Most of the studies on the influence of chemicals on regeneration provide little information concerning the actual biochemical processes, since the action of these compounds is almost certainly indirect (antimitotics, ions, etc.). The possible exception is the study of Flickinger ('59) who was able to reverse polarity of transverse body sections of Planaria, and even of intact whole worms, by arresting protein synthesis in anterior halves of the specimens; however, even in these cases, reversal of polarity was achieved only if the preparations were treated with both

puromycin and colcemide. Since colchicine, a closely related compound, had been found to reverse polarity by itself (McWhinnie, '55), the necessity to use colcemide was unfortunate for interpretation of Flickinger's experiments: it increased the uncertainty as to whether the organization of anatomical structures in planaria depends mainly on the gradient of protein synthesis, a postulate offered by the author. Lindh ('57) studied changes in DNA, RNA, and free nucleotides during regeneration of *Euplanaria polychroa*, but made no attempt to correlate them with particular inductions or inhibitions occurring during this process. Gabriel ('65) found that β-mercaptothanol stopped differentiation, but not the formation of blastemas in isolated posterior body halves, while actinomycin B tended to inhibit both. The degree of inhibition of regeneration after actinomycin was proportional to the decrease in bashophilia of nucleoli in neoblasts of experimental animals.

Recently a hypothesis that in lower vertebrates activation of specific genes is involved in embryonic induction has been explored (Barth, '65). Some evidence that genes may influence primary embryonic induction in amphibia was recently supplied by Ansevin, 1969a; however the problem is by no means clear (Ansevin, '65). Transcription of RNA, as evidence of the gene action, should be investigated in a variety of apparent inductions in various animal groups. This might provide information as to whether transcription is involved at all in induction, and also whether the mechanisms of the so-called inductions are indeed similar in all the cases. The present study on the effects of actinomycin on regeneration of postpharyngeal body sections of *Dugesia tigrina* was undertaken to obtain such information.

MATERIAL AND METHODS

Specimens of *Dugesia tigrina*, 8–10 mm in length, were starved for at least one week before sectioning to lessen the chances of infection from the contents of the intestine. They were then anaesthetized by placing them on a piece of wet filter paper on a petri dish half filled with ice and covered with aluminum foil. A relaxed condition of the worms allowed for more consistency in cutting sections of uniform length. The first cut was made at a manageable distance below the worm's pharynx, and the second cut was then performed a pharyngeal length further down the worm (fig. 1). This was done under the dissecting microscope with two fine needles ("minuten needle," Clay-Adams).

The dosage of actinomycin D [1] was selected for each sample before the experiment was begun by subjecting groups of worms to varying concentrations of the inhibitor. Sublethal concentrations, or concentrations which were lethal after 7–10 days, were used; these appeared to be different for different periods of regeneration as well as for different samples of the antibiotic. Thus, sections treated immediately after the operation could not survive doses above 10 units/ml, whereas fragments which were allowed to regenerate for 24 hours or longer before treatment could withstand doses up to 30 units/ml. Since actinomycin is rather insoluble in water, samples were dissolved in 100% alcohol and then diluted with enough air-bubbled water to produce the desired concentration. Corresponding control sections in these experiments were always treated with the same concentration of alcohol in water as was used for making the respective solutions of actinomycin.

The procedures followed in this investigation are shown in figure 1. In some instances the isolated postpharyngeal sections were immediately transferred to solutions of actinomycin D. In others, the sections were first allowed to regenerate in water for varying intervals of time (from 1 to 144 hours), and were then subjected to the action of actinomycin. In another series, whole worms were kept in actinomycin before sectioning. Both in the sections immediately placed in actinomycin and in those allowed prior regeneration in water, the actinomycin treatment lasted for 24–48 hours; it was followed by thorough rinsing of the sections with water and transfer to new dishes in which air-bubbled tap water was changed every second day. Treatment with actinomycin was always performed in the dark.

The regenerating fragments were maintained up to two or three weeks, with frequent microscopic examinations. Interesting specimens were either dissected or fixed in Bouin fluid or Carnoys fixative, and mounted *in toto*. Altogether, regeneration of 790 transverse postpharyngeal sections of *Dugesia tigrina* was followed in this investigation; 610 sections had been treated with actinomycin D and 180 were controls.

Most of these experiments were performed with worms which came from a colony previously maintained in the laboratory for two years. This colony became exhausted before all the experiments had been completed. A new heterogenous colony was established and an attempt was made to complete the experiments with specimens from this population. However, both control and actinomycin-treated specimens from the new colony appeared to have different properties from those characteristic for the members of the old colony. It was clearly impossible to consider and analyze both populations of *Dugesia tigrina* as one uniform group; therefore, the results obtained from experiments with each population will be described separately.

In order to determine whether actinomycin D indeed suppressed synthesis of new

[1] Courtesy of Mr. Walter A. Gall, Merck Sharp and Dohme Research Laboratories, Rahway, New Jersey.

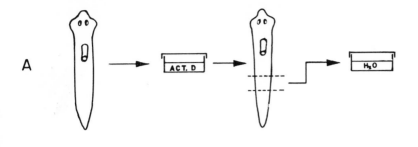

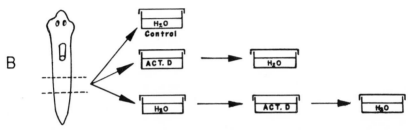

Fig. 1   Design of the experiments: (A) Treatment of whole specimens with actinomycin D from 24 to 0 hours *before* the operation; (B) Treatment of isolated postpharyngeal sections either immediately after the operation or after a time lag. Control sections are allowed to regenerate in water throughout the experiment.

RNA, the following test was performed. Sixty large specimens of *Dugesia* were cut into viable fragments; half of this material was placed in water, the other half was treated with actinomycin at a concentration of 20 units/ml. After four hours 10μC of ^{32}P (as sodium phosphate) was added to each sample. After 24 hours the samples were thoroughly washed with water and homogenized in 10 ml of cold 0.5 N perchloric acid (PCA). Following this the samples were prepared for determination of their contents of RNA and of its specific activity. The procedure is outlined below.

(1) Samples were left in cold PCA for 30 minutes for precipitation to occur, then (2) they were centrifuged for ten minutes at 8000 rpm in Sorval RC-2 4·25 rotor; (3) supernatant solution was discarded, the pellet resuspended in 10 ml of 0.5N cold PCA, and centrifuged; (4) procedure 3 was repeated; (5) the pellet was dried with a kleenex, mixed with 2 ml of 0.3 M KOH, and incubated at 37°C for 18 hours; (6) after chilling the samples, 0.36 ml of cold 5 N PCA was added to each of them, and they were centrifuged for ten minutes at 8000 rpm; (7) supernatant solution was recovered and the pellets discarded; KOH was added to bring pH to 10.0; (8) the samples were again centrifuged for ten minutes at 8000 rpm and the supernatant solution recovered; (9) optical density of the samples was measured in Gilford 2000 model spectrophotometer; (10) the samples were diluted ten times and their specific activity measured in the liquid scintillation counter (Nuclear-Chicago Unilux II).

RESULTS

Specific activities of non-treated and actinomycin-treated regenerating body fragments which were incubated with ^{32}P

79

were 14,996 cpm and 995 cpm, respectively. This indicates that after a four-hour exposure to actinomycin synthesis of total RNA was suppressed by more than 90% of that which is normal.

I. *Dugesia tigrina from the "old" colony*

1. *Regeneration in control sections.* Of the isolated postpharyngeal sections, about 97% survived and regenerated into normal specimens. Head and tail blastemas could be detected on the second or third day after the operation. Mouth opening regenerated by the fourth or fifth day, and the pharynx appeared between the fifth and the ninth day. In 14.8% of the regenerated controls, a second pharynx formed simultaneously; it was most frequently positioned in a transverse orientation to the main body axis, rarely parallel or reverse in relation to the main pharynx. Eyespots usually appeared by the ninth to twelfth day.

2. *Regeneration in sections treated with Actinomycin D.* Of 405 sections which were treated with actinomycin, 376 (92.8%) survived at least long enough to accomplish a major part of the regenerative process; 37% of the regenerates died, but 63% recovered completely. Virtually all treated specimens showed a more or less severe syndrome of symptoms: reduced activity, extensive cytolysis of internal tissues, and considerable swelling of the body.

Data concerning regeneration of the head and pharynx in control and actinomycin-treated postpharyngeal sections are summarized in tables 1 and 2. It is evident from table 1 that no distinct correlation could be found between the stage of regeneration at which this treatment was administered and defective regeneration of the head: although there was appreciable delay in some instances, anterior blastemas formed in the majority of specimens in each experimental group, and most differentiated further into normal heads. Even more confusing was the fact that in some instances a whole sample of worms either failed or succeeded in regenerating heads, while another sample after a similar treatment gave an intermediate or opposite result. However, exposure to the inhibitor during the first 24 hours after the operation generally interfered more consistently with head regeneration than did treatment at later times. This paralleled the effect of the antibiotic on establishment of polarity in the regenerating sections: several bipolar specimens or worms with distorted polarity that developed from postpharyngeal fragments (fig. 2) belonged to the experimental groups which had received the

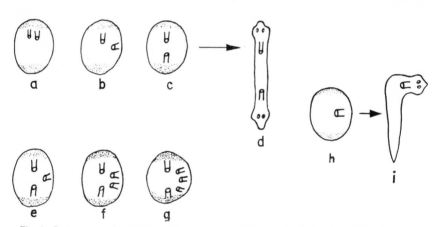

Fig. 2 Positions of supernumerary pharynges in multipharyngeal regenerates resulting from treatment with actinomycin D at early stages of regeneration. Note modifications of polarity of the body in some cases of the multipharyngeal condition.

80

actinomycin treatment within the first 24 hours of regeneration. An exception was a single control bipolar specimen.

Table 2 shows that a definite correlation existed between the number of regenerates which developed as multipharyngeal and the stage of regeneration at which they had been exposed to actinomycin. The treatment was most effective if it was administered before the onset of regeneration or during the first 24 hours; the effect gradually decreased to the control value if treatment was delayed until the third day. However, unlike the bipharyngeal controls, some actinomycin-treated regenerates developed as many as three, four, and, in one instance, even five pharynges.

As is evident in figure 2, the supernumerary organs could lie parallelly, transversely or reversely to the main pharynx; in some cases, the position of the head was also modified according to the orientation of the supernumerary pharynx, or a second head developed in front of it.

Pharynges in actinomycin-treated sections differentiated at the same time as in the controls, i.e., between five and eight days after regeneration. Because of body swelling, it could not always be determined whether supernumerary pharynges differentiated simultaneously with the main one; this was detected in several cases.

II. Dugesia tigrina from the "new" colony

Incidence of multipharyngeal regenerates in control and actinomycin-treated worms of the "new" colony.

The results in this series of experiments are represented in table 3.

Among the 84 control postpharyngeal sections of Dugesia from the newly established laboratory colony, only one bipharyngeal regenerate developed (1.19%).

Of 153 surviving experimental sections which had been treated with actinomycin starting at 1, 2, 3, 4, 5, or 6 hours after regeneration only 7 specimens developed a second pharynx (4.56%). No truly bipolar forms nor a higher number of pharynges were observed in any individual of this experimental group, despite the fact that a number of worms from this population were originally found to be bipharyngeal. All of 15 postpharyngeal sections isolated from such bipharyngeal individuals regenerated only a single pharynx.

DISCUSSION

Comparison of regeneration in control postpharyngeal sections of worms from two laboratory colonies shows that the species of Dugesia tigrina includes strains in which control of regeneraton of a single pharynx is unstable. In one strain, the second organ usually lay transversely or reversely; in the

TABLE 1

Regeneration of the head in control and actinomycin-treated sections from the "old" colony

Time in hours (in relation to the onset of regeneration) at which the actinomycin treatment began	Total number of regenerates	Blastemas failed to form number of specimens	Blastemas formed, but did not further differentiate (number of specimens)	Head structures regenerated (number of specimens)
−24	10	—	—	10
0	32	2	3	27
1	8	2	2	2
2	9	2	2	5
3	8	2	2	4
4	9	1	1	7
6	20	7	9	4
7	17	4	4	9
24	27	—	1	26
48	27	2	8	17
72	39	9	9	21
96	15	—	—	15
120	19	—	—	19
144	5	—	—	5
Controls	96	2	—	94

TABLE 2

Regeneration of the pharynx in actinomycin-treated and control postpharyngeal sections from the "old" colony

Time in hours (in relation to the onset of regeneration) at which treatment with Act. D was initiated	Number of relative "survivors"/ total number of treated sections	Number of monopharyngeal regenerates	Monopharyngeal regenerates	Number of bipharyngeal regenerates	Bipharyngeal regenerates	Number of tri- (or more) pharyngeal regenerates	Tri- (or more) pharyngeal regenerates	Total multipharyngeal regenerates	Number of days of regeneration of the pharynx
			%		%		%	%	
−24	19/30	9	47.4	10	52.6	0	0	52.6	5–7
0	69/71	36 [1]	52.2	26	37.7	5	7.3	45.0	5–9
1 hour	9/10	4	44.4	5	55.6	0	0	55.6	5–9
2 hours	7/10	3	42.8	4	57.2	0	0	57.2	5–9
3 hours	9/10	2	22.2	6	66.7	1	11.1	77.8	5–9
4 hours	10/10	8	80	2	20.0	0	0	20.0	5–9
5 hours	9/10	2	22.2	6	66.7	1	11.1	77.8	5–9
6 hours	10/10	7	70.0	2	20.0	1 [2]	10.0	30.0	5–9
7 hours	52/55	25 [1]	48.0	17	32.6	9 [3]	17.0	49.6	5–9
24 hours	46/50	29	63	15	32.5	2 [2]	4.5	37.0	5–9
48 hours	44/50	32 [1]	72	10	22.7	1	2.3	25.0	5–9
72 hours	43/45	36	83	5	11.6	2	4.7	15.3	5–9
96 hours	29/35	25 [1]	86.2	2	6.9	0	0	6.9	5–9
Controls	81/83	69	85.2	12	14.8	0	0	14.8	5–7

[1] In one or a few specimens the presence of a pharynx could not be ascertained due to an extreme swelling of the body.
[2] One specimen in this group had four pharynges.
[3] One specimen in this group had five pharynges; the remaining eight regenerates were tripharyngeal.

82

TABLE 3

Incidence of multipharyngeal regenerates which developed from postpharyngeal body sections of worms from the "new" colony with or without treatment with actinomycin D

Number of hours of regeneration before the beginning of the actinomycin treatment	Total number of surviving sections in this group	Number of regenerates with 1 pharynx	Number of regenerates with 2 pharynges	Number of regenerates with no pharynx	Number of regenerates with one transverse pharynx	Average percent of bipharyngeal regenerates
0 hour	5	4	1	0	0	
1 hour	34	30	3	1	0	
2 hours	23	21	0	1	1	
3 hours	23	22	1	0	0	4.56
4 hours	13	12	1	0	0	
5 hours	23	12	0	10	0	
6 hours	22	22	1	0	1	
Controls	84	80	1	1	2	1.19

latter case, it sometimes was accompanied by a second, reversely oriented head forming at the posterior apex of the regenerate, thus suggesting a relation between distortion of normal polarity of the organism and induction of supernumerary pharynges. In a different group of worms, another abnormality was detected: some individuals had a second pharynx lying right behind the main one in roughly the same orientation. Stephan ('65) reported instability in the number and position of pharynges in a laboratory strain of *Dugesia tigrina* in which, apparently due to duplication or triplication of the main body axis, two or three pharynges would lie parallelly to each other.

In the present experiments, the abnormality involving unstable polarity and a varying number of pharynges proved useful in studying dependence of pharynx regeneration on synthesis of new RNA in postpharyngeal body sections. However, arrest of RNA synthesis in the sections had a rather unexpected effect: in spite of virtually lethal doses of actinomycin D, pharynges could still be induced and differentiate and, in fact, regenerated in even higher numbers than under normal conditions. This effect was noted only in those regenerates in which synthesis of RNA had been considerably slowed down at any time during the first 48 hours of regeneration, and, actually, was less frequent in individuals in which it had been left undisturbed during the first 24 hours. Exposure to actinomycin at any later stages of regeneration did not affect normal differentiation of a single pharynx. Also, this response to actinomycin was mostly limited to one laboratory strain; the other strain, although including some spontaneously bipharyngeal individuals, gave such a low incidence of multipharyngeal regenerates, both in the controls and in experimental specimens after actinomycin treatment, that no significance can be attached to the results.

It does not seem likely that the effects of actinomycin on regeneration of pharynges and on polarity were directly caused by destruction of the old organization in the section and, thus, only indirectly related to the RNA turnover in the cells. Since the postactinomycin cytolytic syndrome re-

83

sulted in similar internal disorganization of the tissues, no matter at what stage of regeneration the treatment was given, had disorganization been the main factor, reconstruction of the internal organization should have been equally imperfect irrespectively of the time of the treatment. But exposure to the antibiotic after 48 hours of undisturbed regeneration never impaired, although it did delay its final outcome, while early treatments had a distinctive modifying action. Also, the syndrome was as severe in body fragments of worms from the "old" colony as from the "new" population, and yet regeneration was modified only in the former. Furthermore, the test for incorporation of ^{32}P into newly synthetized RNA showed that in the present experimental conditions actinomycin blocked more than 90% of total synthesis of RNA. Since the yield of RNA from planaria is very low and this would make isolation of messenger RNA an exceedingly difficult task, no attempt was made to determine specific activity of the isolated messenger RNA fraction. However, since it is well known that actinomycin specifically arrests nuclear transcription (Harbers, Domagk and Müller, '64), and since the dose used was high, it is unlikely that transcription of informational RNA in the regenerating planarian fragments would have selectively escaped inhibition. Thus, the likelihood is considerable that the observed morphogenetic modifications represented interference with genetic transcription during regeneration. If so, different effects of actinomycin applied at different stages of regeneration would be indicative of sequential control of genes over particular regenerative processes.

An interpretation which we offer to explain the observed effects of actinomycin on regeneration is based on the assumption that, in *Dugesia tigrina*, messenger RNA for coding proteins which transform undifferentiated neoblasts into the tissue of the pharynx exists continuously even in the intact worm. This messenger RNA is, under normal conditions, inactivated by a short-lived inhibitor which possibly diffuses from the existing differentiated pharyngeal zone and is induced in it by the original pharynx (existance of such inhibitor in homogenates from appropriate body

parts was detected by Ziller-Sengel, '65, '67a,b). When this block is removed (by isolation of the postpharyngeal section), the existing pharynx-mRNA is liberated and begins to code pharynx-cell proteins. This happens throughout the body fragment so that actually several pharynges become induced in random orientations at this point (fig. 3a). However, which of these will eventually reach morphological expression, and what will be their positions in the reconstituted worm depend on establishment of new polarity in the regenerating section. This process which happens next apparently requires synthesis of new mRNA, since actinomycin applied between 7 and 24 hours of regeneration causes distortion of polarity in some cases (fig. 2). Establishment of new polarity results in determination of the head and tail regions (fig. 3b) which, while differentiating, synthetize a substance capable of arresting some postinductive reaction in differentiation of the pharynx. This inhibitor would prevent morphological expression of all induced pharynges, if another regenerative event has not prevented it: a "postcerebral zone" is induced by the head (fig. 3c) and becomes the source of limited synthesis of a factor capable of neutralizing the head inhibitor. However, in the region of the head, the concentration of the latter is too high so that not all of it becomes inactivated, and thus, the pharynx never differentiates in the head region. Conversely, below the postcerebral zone the amounts of the inhibitor are smaller and all of it can be blocked by the "neutralizer" which provides for differentiation of a pharynx to proceed unhindered in this zone. Below the pharyngeal area the "neutralizer molecules" become scanty, while more inhibitor is supplied from the differentiating tail; therefore, completion of differentiation of a pharynx is again prevented. Thus, in the light of the proposed model, reconstitution of the whole organism from a body fragment is the outcome of a succession of local inductions and inhibitions.

This working model is actually in good general agreement with the earlier scheme of Wolff ('53) which, for comparison, is also represented in figure 3 d, e, f. There are two differences: Wolff assumes that the postcerebral zone becomes established

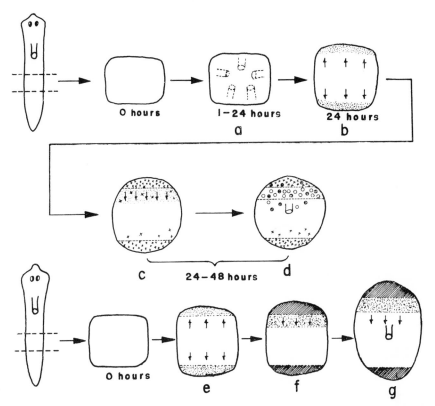

Fig. 3 Diagrammatic representation of induction and inhibition taking place early in morphallactic regeneration of the postpharyngeal section according to: (A) the proposed hypothesis; (B) Wolff's hypothesis.

first in regeneration and is then an inductor for the pharyngeal area; whereas we suggest that induction of pharynges precedes establishment of the postcerebral zone. Also, Wolff considers induction of only a single pharynx, while we suggest that several pharynges at random orientations become determined initially. The observed differences between the strains in their normal incidence of bipharyngeal or multipharyngeal specimens and in their response to actinomycin could be explained by assuming that particular strains of *Dugesia* may differ in the rate of synthesis of mRNA for differentiation of the pharynges, or of mRNA for establishment of polarity, or of both. For example, the worms of the first strain used in this study, when uninjured, might have an almost subthreshold concentration of mRNA for establishment of polarity, but considerable quantities of pharynx mRNA in their tissues. Hence, under the action of actinomycin, body fragments from this strain would show an increased number of multipharyngeal individuals and a higher incidence of cases of distorted polarity. This is what was actually observed. A strain which would have polarity-mRNA in concentration above the threshold value would show

85

few cases of abnormal polarity and provide few multipharyngeal individuals, both in normal regeneration and after actinomycin, while that in which pharynx-mRNA concentration was low would, after actinomycin treatment, give a higher percentage of regenerates which would lack pharynx altogether or develop it after a delay.

At the present time, it is not possible to create a model of regenerative determinations which would avoid being speculative to a considerable degree. However, we believe that a number of facts support the existence of the developmental factors proposed in our hypothesis, with the exception of the "postcerebral neutralizer" which is still only a purely speculative point. First, pharynx inhibitors have been found to be real in the pharyngeal zone (Ziller-Sengel, '67a), and there are indications (Ansevin, '69b) that they are also present in the differentiating head blastema. That a pharynx never forms next to the head or to a differentiated tail suggests the possibility that another inhibitor may be present in the tail region. Second, numerous cases of supernumerary pharynges were observed in our experiments, but no cases in which abnormal polarity would not have been accompanied by an additional pharynx, or, at least, by a shift in position of the main one. This, in our opinion, speaks for independence of induction of the pharynx in the morphallactically regenerating postpharyngeal section from establishment of a final polarity of the regenerate, and suggests that it precedes the latter. Third, the fact that when, apparently due to the inhibition of synthesis of new RNA, the axial control is not exerted on the regenerating body fragment at an appropriate time, multiple pharynges develop is suggestive that more than one pharynx becomes induced at that early stage of regeneration.

A general impression emerging from these observations is that, in regenerative reorganization of planarian body fragments, previously synthesized long-lived messengers determine to a large extent the new organization in a succession of local "inductions" and inhibitions. The amounts of these messengers are in some cases high enough that even when synthesis of new RNA is substantially cut down, regeneration can be accomplished in a perfectly normal fashion. Thus, what is called "induction" in planaria may be a process which is not primarily dependent on activation of new genes.

The possibility that messenger RNA's for special cell types are present in the cells long before the actual process of differentiation was suggested by other studies. In *Fundulus* embryos (Wilde, '67) in which synthesis of new RNA was irreversibly blocked at early cleavage stages, cells could still differentiate at the proper time, although they were unable to organize into tissues. Ansevin ('65) found that normal induction occurred in frog ectoderm which was kept before, and during induction in the presence of barely sublethal doses of actinomycin. The question of the point in the process of differentiation at which transcription of the messenger is essential clearly remains open.

ACKNOWLEDGMENTS

The authors greatly appreciate the help of Dr. Roger Storck in preparation of the RNA samples and determination of their specific activities. They also thank Mrs. Elga M. Lewis for her expedient and enthusiastic technical assistance.

LITERATURE CITED

Ansevin, K. D. 1965 The effects of RNA synthesis inhibition and of protein synthesis inhibition on embryonic induction and cell differentiation in *Rana pipiens*. J. Morph., *117:* 171–183.
———— 1969a The influence of the nucleus on sequential determination in frog ectoderm following induction by lithium ion. J. Embryol. exper. Morphol., *21:* 383–390.
———— 1969b The influence of a head graft on regeneration of the isolated postpharyngeal body of *Dugesia tigrina*. J. Exp. Zool., in press.
Barth, L. G. 1965 The nature of the action of ions as inductors. Biol. Bull., *129:* 471–481.
Flickinger, R. A. 1959 A gradient of protein synthesis in planaria and reversal of axial polarity of regenerates. Growth, 23: 251–271.
Gabriel, A. 1965 Effects du B-Mercaptoéthanol et de l'actinomycin B sur la régénération de la Planaria *Dugesia gonocephala*. In: Proc. Regeneration in Animals. North Holl. Publ. Co., Amsterdam, pp. 149–159.
Harbers, E., G. F. Domagk and W. Müller 1964 Introduction to Nucleic Acids. Reinhold Book Co., New York.
Kido, T. 1952 Transplantation of planarian pieces divided into dorsal and ventral tissues. Annot. Zool. Japan, *25:* 383–387.
Lender, Th. 1952 Le rôle inducteur de cerveau dans la régénération des yeux d'une planaire d'eau douce. Bull. Biol. France, 85: 140–215.

Lender, T., and P. Gripon 1962 La régénération des yeux et du cerveau du *Dugesia lugubris* en présence de deux tronces nerveaux inégguax. Bull. Soc. Zool. France, 87: 87–95.

Lindh, N. O. 1957 The nucleic acid composition and nucleotide content during regeneration in the flatworm *Euplanaria polychroa*. Arch. f. Zool., 11: 153–166.

McWhinnie, M. 1955 The effects of colchicine on reconstitutional development in *Dugesia dorotocephala*. Biol. Bull., 108: 54–65.

Miller, J. A. 1938 Studies on heteroplastic transplantations in Tricads; Cephalic grafts between *Euplanaria dorotocephala* and *E. tigrina*. Physiol. Zool., 11: 214–247.

Okada, Y. U., and T. Kido 1943 Further experiments on transplantation in Planaria. J. Fac. Sci. Univ. Tokyo IV., 6: 1–23.

Santos, F. V. 1931 Studies on transplantation in *Planaria dorotocephala* and *Planaria maculata*. Physiol. Zool., 4: 111–164.

Stephan, F. 1965 Régénération des Formes Doubles de *Dugesia tigrina* (Triclade d'Eau Douce). In: Proceedings Regeneration in Animals. North Holland Publishing Company, Amsterdam, pp. 202–206.

Teshirogi, W., and A. Jin 1964 The relation between regeneration and the nervous system in the freshwater planarian, *Bdellocephala brunnea*. I. Regeneration by a cut-piece lacking the ventral nerve cords. Zool. Mag.,73: 45–51.

Török, L. J., and I. Törö 1962 Beitrage sur Problem einer "morphogenetischen Hemmung" an Hand experimenteller Befunde dei der Regeneration von Planarian. Embryologia, suppl., 6: 319–354.

Wilde, C. E., Jr. 1967 Epithelial-Mesenchymal Interactions in the Lower Vertebrates. 18th Hahnemann Symp: Epithelial-Mesenchymal Interactions, Philadelphia.

Wolff, E. 1953 Les phénomenes d'induction dans la régénération des Planaires d'eau douce. Rev. suisse Zool., 60: 540–546.

Ziller-Sengel, C. 1965 Inhibition de la Régénération du Pharynx chez les Planaires. In: Proceedings Regeneration in Animals. North Holland Publishing Company, Amsterdam, pp. 193–201.

——— 1967a Recherches sur l'inhibition de la régénération du pharynx cher les planaires. I. Mise en évidence d'in facteur auto-inhibiteur de la régénération du pharynx. J. Embryol. exp. Morphology, 18: 91–105.

——— 1967b Recherches sur l'inhibition de la régénération du pharynx chez les planaires. II. Variations d'intensité du facteur inhibiteur suivant les especes et les phases de la régénération. J. Embryol. exp. Morphology, 18: 107–119.

1-2 Two of the bipharyngeal specimens which developed with high frequency from postpharyngeal sections treated with actinomycin D at early stages of regeneration. Worm 2 is not fully bipolar.

3 One of the two four-pharyngeal specimens which developed in the same experimental group: this worm was dissected, and during fixation two of the pharynges (indicated by arrows) became detached and displaced to a position above the specimen.

PLATE 1

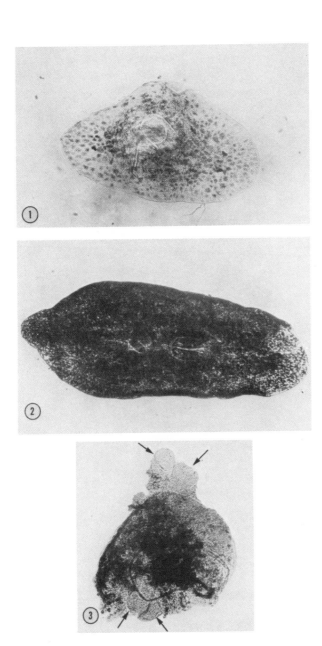

Pathology of Turbellaria and
Effects of Radiation

Observations on some tumours found in two species of planaria—*Dugesia etrusca* and *D. ilvana*

By C. S. LANGE

Since the description by Pallas in 1774 of their remarkable capacity for regeneration the planarians have been of interest to the biologist (cf. Brøndsted's excellent review—Brøndsted, 1955). Dubois (1949) has shown that regeneration in planaria is due to undifferentiated totipotential cells referred to as neoblasts, which are the only cells known to divide mitotically. When the animal is cut or similarly injured, some neoblasts in the region of the wound rapidly differentiate to form a new epidermis, while other neoblasts divide and migrate towards the wound surface where they continue to divide and/or differentiate into various cell types. The finding of spontaneous growths or tumours in this simple material (Goldsmith, 1939) suggests that it may be of interest to study the nature of such tumours.

MATERIALS AND METHODS

Material for the present studies includes the species *Dugesia etrusca* and *D. ilvana* obtained from the laboratory of Professor Mario Benazzi of the University of Pisa, Italy. *D. etrusca* was collected in Toscana and *D. ilvana* from Isola d'Elba. Both species are diploid. The animals, ten of each species, were kept individually or in pairs in 13 ml. polystyrene tubes (Camlab) filled with tap water and maintained at 20 ± 1 °C. They were fed on washed slices of liver (beef, sometimes lamb) twice weekly. The water was changed, and waste matter removed, 24 h after feeding. Under these conditions both species multiply rapidly by transverse fission. Progeny were placed in separate tubes as often as was necessary to prevent overcrowding. After 17 weeks the ten animals of *D. etrusca* and *D. ilvana* had multiplied to ninety-four and eighty-eight respectively, and at that time eighteen and twelve animals, respectively, were found to have tumorous growths (Text-fig. 1). The animals were therefore isolated and the progress of the tumours was followed. Weekly checking of the normal population showed further examples of such tumours. A total of fifty-four tumour-bearing animals was studied. No difference was observed between the tumours of

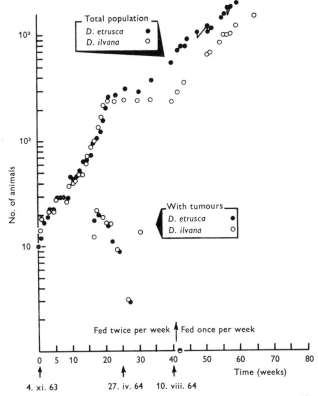

Text-fig. 1. Rates of population increase and appearance of growths.

D. etrusca and *D. ilvana*, and they can be described together. *D. ilvana* is indistinguishable from *D. etrusca* on the basis of body shape (Plate 1, fig. A).

OBSERVATIONS

Most of the tumours began in the posterior tip, appearing as a reddened (normal pigmentation is a tan or slightly red-tan) swelling or lump or group of lumps which stretched the skin and distorted the shape of the tail (Plate 1, fig. B). The tail became very much thicker and impeded the planarian's movement. The lumps grew slowly, became darkly pigmented, and eventually developed clear, white, tubes from which mucus was at times expelled (Plate 1, fig. C).

The planarians, which were rapidly growing and increasing in number, continued to divide even after the tumour had become the largest part of the body. The anteriors gave rise to healthy animals in which recurrence of the tumours has not yet been observed (1 year); while the posterior portion retained the

PLATE 1

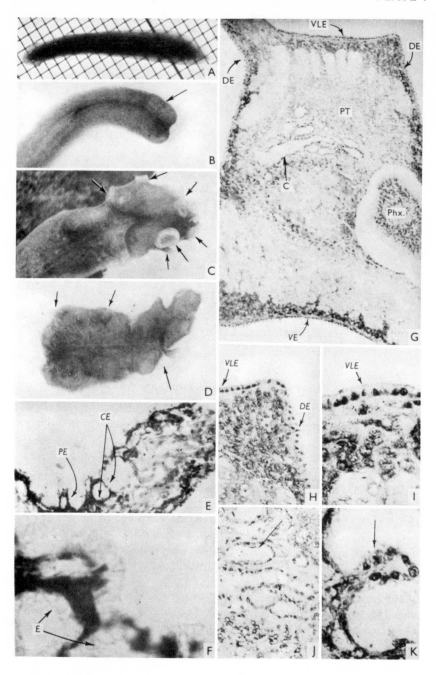

tumour. The posterior portion regenerated a new head at the normal rate and divided again, leaving a smaller posterior piece containing the tumour (Plate 1, fig. D). Sometimes the second division took place before the 'posterior' head was completely regenerated; however, this occasionally happened in the normal animal as well. Finally, the posterior portion as well as the tumour lysed.

In one case (*D. etrusca*) a planarian was found with growths on both the anterior and posterior ends. No fission occurred and the mid-region shrank rapidly until final lysis of both this region and the tumours occurred.

After cutting, normal tissue is regenerated whereas the tumour tissue is not. After a cut through normal tissue, the planarian's muscles contract, pinching off the wounded area and thus preventing cells from pouring out into the surrounding environment. Local neoblasts differentiate to form a new epithelium over the wound surface. Finally, a stimulation occurs in other nests of neoblasts as more cells are needed, and the normal regenerative phenomena follow. However, after a cut through the tumour, the pinching off of the wounded area did not occur. Cells streamed out until all the tumour tissue, as well as a large quantity of normal tissue, had leaked out. Then the pinching off occurred, well beyond the original boundaries of the tumour, and normal regeneration followed.

When the animals were reduced to one feeding per week no new occurrences of tumours were found in the population as a whole in my material (apart from two examples which are described below and which were first observed one week after feeding had been reduced to once a week).

In two cases, the tumour started in the mid-region of the dorsum just posterior to the mouth and assumed a wide crater-like shape with a clear ring of tissue forming a rim around the apex of the mound. In both cases the planarians did not divide and more tumours appeared on the posteriors, which soon afterwards began to lyse. At this point both animals were fixed for histological study. Morphologically, the crater-like tumours appeared very much like the advanced stages of the other tumours described, only being somewhat larger and perhaps more advanced. Their position immediately posterior to the pharynx may be the reason for the anteriors not dividing away from the tumours (fission normally takes place through this region).

Histological preparations of tumour-bearing animals show an enlarged tail thickness, consisting mainly of papillary and crypt-like structures covered and lined, respectively, by mucus-producing epithelium (Plate 1, figs. E, F); and a broken or perforated basement membrane. The epithelial lining of the crypt-like structures is typical of most of the tumours described at the beginning of this paper.

The crater-like tumour is notable for its large mass, the high degree of differentiation of some of its tissues (e.g. nervous, parenchymal and epithelial tissues), and its epithelium-lined crypts (Plate 1, figs. J, K). Plate 1, fig. G, shows such a tumour which occurred in the mid-region of the animal just posterior to the mouth; the posterior portion had already lysed (the tumour there was more like the first-described tumours). The top of the tumour is lined with a ventral-like epithelium (Plate 1, figs. H, I).

In methyl-green pyronin-stained sections, large pockets of neoblasts sprinkled about the tumour are visible. At higher magnifications of an epithelium-lined crypt, the epithelial lining is not unlike the ventral epithelium (Plate 1, figs. H, I, J, K), and the ventral-like epithelium-lined tumour surface may have been derived from the opening-up (bursting?) of such a crypt. This might be a similar phenomenon to that observed in embryos of *Polycelis tenuis* (Skaer, 1965), where vesicles of ciliated cells migrate to the ventrum and open to form the ventral epithelium.

The cells and tissues found in these tumours do not appear to be abnormal in themselves, but rather in their position with respect to their surrounding tissues and the rest of the animal.

CONCLUSIONS AND DISCUSSION

Other workers have reported abnormal growths in planarians. Melander's description (Melander, 1950) of growths associated with the occurrence of accessory chromosomes in *Polycelis tenuis* is totally unlike the tumours which

---

PLATE 1

Fig. A. Normal planarian (*Dugesia ilvana*). Each square is 0·9 mm a side. Posterior on right, anterior on left. (× 2·7.)

Fig. B. Posterior portion of a tumour-bearing planarian. This is the first visible stage in the development of the growth. The arrow shows the dorsal epithelium stretched over the lump. (× 3·3.)

Fig. C. Differentiated stage of the growth. The clear white tubes are indicated by the arrows. (× 4·8.)

Fig. D. Final stage of the growth. Rapid, complete lysis occurs shortly after this stage. The arrows show that the growth (tubes) has reached the anterior edge (left) of this small posterior piece. (× 5.)

Fig. E. Posterior portion of a sagittal section of a tumour-bearing planarian. Methyl-green pyronin, × 204. *CE*, Epithelium-lined crypt; *PE*, epithelium-lined papillary structure. Dorsum on upper left, ventrum on lower right corner.

Fig. F. Enlargement of part of fig. E (× 892). *E*, Epithelial lining of crypt.

Fig. G. Sagittal section of the crater-like growth, showing the differentiation of the tissues, i.e. the epithelium-lined crypts (*C*), parenchymal tissue (*PT*), the dorsal (*DE*) and ventral-like (*VLE*) epithelia covering, respectively, the sides and top of the tumour. Haematoxylin, × 112. *Phx.*, Pharynx; *VE*, ventral epithelium.

Fig. H. Enlargement of part of fig. G, showing the dorsal epithelium (*DE*) on the anterior facing rise of the growth and the ventral-like epithelium (*VLE*) covering the top (dorsal surface) of the growth. Haematoxylin, × 254.

Fig. I. A further enlargement, showing the ventral-like epithelium (*VLE*) covering the dorsal-facing top of the tumour. Compare with fig. K. Haematoxylin, × 605.

Fig. J. The epithelium-lined crypts. Haematoxylin, × 245. Compare with fig. H.

Fig. K. Higher magnification, showing part of an epithelium-lined crypt. Compare with Fig. I. Haematoxylin, × 617.

I have observed. Stéphan (1960) gives a detailed description of tumours in *D. tigrina*, first described by Goldsmith (1939, 1941), which although morphologically similar in many ways to the tumours found in *D. etrusca* and *D. ilvana*, differ strikingly from them. In *D. tigrina*, tumours recurred with a high frequency. Furthermore, in no case of mine did the tumours differentiate into heads or tails, as reported by Stéphan (1960) for *D. tigrina*.

D. etrusca and *ilvana* are localized species of the '*gonocephala*' group and are classified as *D. etrusca* Benazzi and *D. ilvana* Benazzi.

The literature also contains several reports of induction of planarian tumours, but Foster (1963) is the only one to suggest that these growths are neoplastic. It is also possible that these tumours are the result of faulty differentiation control—a not unlikely possibility, as differentiation of the totipotential neoblast is determined by the surrounding non-mitotic somatic tissue (Dubois, 1949; Lender, 1960; Kolmayer & Stéphan-Dubois, 1960; Stéphan-Dubois, 1961; Stéphan-Dubois & Gilgenkrantz, 1961).

A further study of growths of this kind may be particularly instructive on aspects of oncogenesis not seen in animals with a low capacity of regeneration.

SUMMARY

Spontaneously differentiated tumours in planaria which had been multiplying rapidly by transverse fission are described. Development was progressive, leading to lysis of both tumour and host. No recurrence was observed in anteriors of animals which had divided away from tumour-bearing posteriors, nor were new occurrences observed in slower-growing animals fed once, instead of twice, per week. It is suggested that the tumours might be the result of differentiation error rather than neoplasia.

The author wishes to thank Drs S. Muldal, L. G. Lajtha and Alma Howard for helpful discussions, the Medical Illustration Department of the Christie Hospital and Holt Radium Institute for help in preparing the illustrations, and Professor M. Benazzi of the University of Pisa for providing the original ten specimens of each of the two species.

REFERENCES

BRØNDSTED, H. V. (1955). Planarian Regeneration. *Biol. Rev.* **30**, 65–126.
DUBOIS, F. (1949). Contribution à l'étude de la migration des cellules de régénération chez les Planaires dulcicoles. *Bull. biol. Fr. Belg.* **83**, 215–83.
FOSTER, J. A. (1963). Induction of neoplasms in planarians with carcinogens. *Cancer Res.* **23**, 300–3.
GOLDSMITH, E. D. (1939). Spontaneous outgrowths in *Dugesia tigrina* (Syn. *Planaria maculata*). *Anat. Rec.* **75** (suppl.), 158–9.
GOLDSMITH, E. D. (1941). Further observations of supernumerary structures in individuals of an artificially produced clone of *Dugesia tigrina*. *Anat. Rec.* **81** (suppl.), 108–9.
KOLMAYER, S. & STÉPHAN-DUBOIS, F. (1960). Néoblasts et limitation du pouvoir de régénération céphalique chez la Planaire *Dendrocoelum lacteum*. *J. Embryol. exp. Morph.* **8**, 376–86.

LENDER, TH. (1960). L'Inhibition spécifique de la différenciation du cerveau des Planaires d'eau douce en régénération. *J. Embryol. exp. Morph.* **8**, 291–301.

MELANDER, Y. (1950). Accessory chromosomes in animals, especially in *Polycelis tenuis. Hereditas,* **36**, 19–38.

SKAER, R. J. (1965). The origin and continuous replacement of epidermal cells in the planarian *Polycelis tenuis* (Ijima). *J. Embryol. exp. Morph.* **13**, 129–39.

STÉPHAN, F. (1960). Tumeurs spontanées chez la Planaire *Dugesia tigrina. C. r. Séanc. Soc. Biol.* **156**, 920–2.

STÉPHAN-DUBOIS, F. (1961). Les cellules de régénération chez la Planaire *Dendrocoelum lacteum. Bull. Soc. zool. Fr.* **86**, 172–85.

STÉPHAN-DUBOIS, F. & GILGENKRANTZ, F. (1961). Transplantation et régénération chez la Planaire *Dendrocoelum lacteum. J. Embryol. exp. Morph.* **9**, 642–9.

A POSSIBLE EXPLANATION IN CELLULAR TERMS
OF THE PHYSIOLOGICAL AGEING OF THE PLANARIAN

C. S. LANGE

INTRODUCTION

ABELOOS (1930) once said that a cell which continually divides is capable of immortality, whereas a cell which has ceased to divide is fatally committed to senility and death. Developing this theme to the level of the organism, he stated that in planarians there are stem cells which, through periods of tissue renewal and regeneration, are able to control ageing; but that even in planarians, once growth has ceased, normal tissue renewal is insufficient to assure the permanent equilibrium (between young and old cells) of the somatic tissues.

In a quantitative study of the number and distribution of the neoblasts (stem cells) of the planarian *D. lugubris*, Lange (1967) has demonstrated relationships between planarian size and neoblast number, and planarian size and total tissue volume. Under the conditions of that study, planarian size was a function of age. Further data, presented in this paper, interpreted in the light of our present knowledge of the mechanisms involved in planarian growth and regeneration, are thought to demonstrate the validity of Abeloos' statement and to suggest the reason for the inability of normal tissue renewal to prevent ageing of the planarian. Before we examine the data, however, let us consider the following two problems: the cellular basis for planarian regeneration, and what constitutes ageing in the planarian.

THE CELLULAR BASIS FOR PLANARIAN REGENERATION

The suitability of the planarian as a material for the study of regeneration was recognized as early as 1774 when Pallas first described planarian head-regeneration. Since then we have learned much more about the mechanism(s) of regeneration in the planarian (cf. the extensive review by Brøndsted (1955); and Lange, in preparation). Child (1915), amongst others, recognized the rejuvenatory effects of regeneration but did not make further use of the planarian as a material for the experimental study of ageing. Curtis (1928) and Curtis and Schultze (1934) recognized the importance of the planarian neoblast as a stem cell for regeneration and possible daily tissue repairs. More recently, the studies of Dubois (1949), Le Moigne (1963, 1965a, b, c, 1966) and Fedecka-Bruner (1964) have conclusively demonstrated the stem cell nature of the neoblast, while those of Lender (1950, 1952, 1956a, b, c, 1960), Sengel (1951a, b, 1953), Kolmayer and Stéphan-Dubois (1960), and Fedecka-Bruner (1961), showed the importance of the differentiated somatic tissues in the control of neoblast differentiation. For at least two organs (brain and eye), the differentiation control has been shown to be exerted by means of hormones.

Our knowledge to date allows us to consider the planarian to be composed of two cell populations: one, the non-replicating differentiated tissues which carry out the various functions necessary for (planarian) survival, interspersed with the second, a morphologically identifiable population of totipotential stem cells. Survival is then seen to depend on having an adequate supply of stem cells to provide the anlagen for differentiation, and properly functioning differentiated cells to organize the course of the regeneration (both in the blastema and locally throughout the body) and to carry on the functions essential for the survival of the organism. Brøndsted (1956) has shown that the differentiated tissues ("the field") determine the rate of regeneration while the quantity of neoblasts determines the size of the regenerated parts.

A study of the loss of regenerative ability (ageing?) in the planarian should then consider: what is the cause of this loss? Is there a shortage of stem cells, or is the differentiation control at fault, or both?

WHAT CONSTITUTES AGEING IN THE PLANARIAN?

Haranghy and Balázs (1964) briefly reviewed the data on planarian lifespan and concluded that two different definitions of lifespan are needed: one for the sexually reproducting species (moment of hatching until death of the individual) and another for the asexually reproducing species* (fission to fission). Although it may be desirable to differentiate between individual lifespan and clonal lifespan, one must beware of the pitfalls implicit in the assumption that an animal about to undergo fission is old as it is near the end of its lifespan. For both sexual and asexual reproduction, what is at stake at the cellular level is the clonal survival of the stem cells. For the former, all of the differentiated tissues are being replaced; while in the latter, only half. The sexually reproducing species may be clonally propagated by repeated cuttings and regeneration, and some species have alternating periods of sexual and asexual reproduction (e.g. *D. dorotocephala*, q.v. Jenkins and Brown, 1963). Thus, again we see that the natural lifespan of an individual need not bear any relationship to the physiological or the chronological age of its body tissues. Yet, when all of the stem cells of a planarian are radiation-sterilized, the cause of death is the physiological run-down (ageing) of the differentiated somatic tissues through lack of replacement of cells lost by daily wear and tear (Lange, in preparation).

For the purposes of gerontology, it should be remembered that the fission products of an asexual reproduction are *not* new individuals, but only partly new, as part of the body tissues are still those of the parent, and even certain classically conditioned responses and discriminations are partly "remembered" in the regenerate (McConnell, Jacobson and Kimble, 1959). Instrumental learning has also been shown to be retained in the regenerated fission product (Ernhart and Sherrick, 1959; Best and Rubenstein, 1962). Thus, certain environmental experiences are passed on, in part, to the fission product or regenerated offspring, through the "inherited" parent tissues.

---

* For the purposes of this definition, the pseudogamic and parthenogenetic species are considered sexual as the new individual arises from a single cell—a gamete.

The above considerations lead us to suggest that what is most important in the gerontological study of the planarian is the study of tissue continuity, i.e. ageing in terms of the loss of ability of planarial tissues to maintain their structural and functional integrity with increasing chronological age. With this outlook in mind, let us now consider the literature on planarial ageing, regeneration and rejuvenation.

Morphological changes

1. *The development of supernumerary ocelli as a sign of ageing.* Balázs (1962) has shown that some supernumerary ocelli and/or their primordia are already present in the embryo of *D. lugubris* and "continue to form throughout the longest lifespan investigated (2·5 yr). In fully-developed individuals, the percentage of accessory eyes and primordia rises . . . (from 1·92 per cent at hatching) . . . to 63·6–72·6 per cent". Abeloos (1930) reported that accessory eye formation in *D. gonocephala* is independent of the size of the animal but depends on temperature, more being produced at higher temperatures. It is correlated not with chronological age but with cessation of growth. Goldsmith (1939, 1941), Stéphan (1962) and Lange (1966) observed spontaneous and induced tumours in several fissiparous planarian species. The former two authors observed the development of supernumerary eyes in the tumours. These findings were interpreted by Lange (1966) to be indicative not of neoplasia, but of "differentiation error". Ghirardelli and Tasselli (1956) found that specimens of *D. lugubris* with supernumerary eyes regenerated heads with the normal number (2) of eyes after decapitation. In view of Lender's (1956a) demonstration that eye differentiation is mediated by the production and release of a hormone ("organisine") from the brain, accessory eye-formation could be taken as a sign of an "hormone imbalance" which may be associated with ageing.

2. *Other morphological changes.* Pigment disintegration and/or loss accompanied by epithelial distension and ulceration have been described by Abeloos (1930), Hyman (1951) and Balázs and Burg (1962a) as signs of senescence, frequently leading to the disintegration of part or the whole of the animal. Unfortunately, this process is not sufficiently protracted to provide a useful scale of ageing, but happens rather rapidly, resulting in death or in regeneration (and rejuvenation) of the surviving fragment(s). These changes are very much like those accompanying the death or recovery of planarians following LD_{50} doses of X– or β^--irradiation (Lange, unpublished data). Other morphological changes, such as the decrease in relative gut capacity, the increase in height of epidermal epithelial cells, and changes in body proportions, are not signs of senility, but are related only to the absolute size of the animal and hence are features of planarian *growth, not senescence* (Abeloos, 1930).

Physiological changes

1. *Decline of sexual reproductive abilities as a sign of ageing.* Balázs and Burg (1962b) and Haranghy and Balázs (1964) describe the age-dependent decline of cocoon production in *D. lugubris* (B?). After an almost linear rise in cocoons laid per individual per month (slope $= +0·3$), which the authors suggest is due to asynchrony of maturation in their material (and an implied constant laying rate for the mature individuals), over the first 3–11 months of life, a 30 per cent drop occurs and cocoon production appears to remain constant thereafter. The number of embryos per cocoon remains relatively constant from 4–20 months (ages for which data are given), but the percent fertility of the cocoons decreases linearly (slope $= -4·4$ per cent/mo.). Reynoldson, Young and

101

Taylor (1965), however, find no such loss of fertility in *D. lugubris* B* over a period of 2 yr. They have shown, however, that the upper limit of temperature tolerance for sexual reproduction in *D. lugubris* B is between 20° and 23°C. At 23°C only 44 per cent of the cocoons hatched (as opposed to 98 per cent at 20°C). Dubois (1949) found 28°C lethal to *D. lugubris* B. As Balázs and Burg allowed their temperature to fluctuate between 18° and 24°C, heat shock may have been an important factor in decreasing fertility. However, Voigt (1928) and Abeloos (1930), under better conditions of temperature control, also reported decreased cocoon production with age in *Polycelis nigra* and *D. gonocephala*, respectively.

2. *Diminution of growth rate, metabolic rate and regeneration ability.* Haranghy and Balázs (1964) claim that in *D. lugubris* no marked difference is found in the rate and capacity for regeneration between mature and old animals, but the rate of blastema growth and differentiation in young (small, 4-weeks-old) animals was considerably higher than that of mature (10–12 months) and aged (2·5–3 yr) (large) individuals. Similar findings for young and old animals were reported by Lindh (1957*a*, *b*) for *D. lugubris* E.† He showed that neoblast mitotic activity in the blastema starts earlier in young animals (6 months) than in old animals (2 yr). Moreover, "2-yr-old worms had a very low mitotic activity, if any at all, and no blastema formation during the first days of regeneration. Sometimes they did not regenerate but died immediately or after a couple of days. The younger the animal was, the higher its mitotic activity and its power of regeneration." (1957*a*). Haranghy and Balázs (1964) did not observe this mortality even in 4-yr-old *D. lugubris*, but the likely difference in biotype of their material may be the reason for this.

Sharov (1934) found that the regeneration ability (head-frequency) of *Dendrocoelum lacteum* embryos decreased from 25 per cent at stages I and II, to 12 per cent at stage III, to 7 per cent in 1-day-old animals. Thereafter, head-frequency increased (25 per cent at 20 days, 55 per cent at 30 days, 60 per cent at 60 days, 80 per cent at 6·2 mm long) to a maximum of 80 per cent in 6·2 mm long animals, and decreased again as the animals grew larger (75 per cent at 10·2 mm, 60 per cent at 14·5 mm).

Abeloos (1930) has amply demonstrated in *D. gonocephala*, that the diminution in growth rate and regeneration ability are correlated with increased planarian size (not age). Hyman (1919) has shown, in *D. dorotocephala*, that the metabolic rate (oxygen consumption per unit weight) of young animals also declines with increase in size. These changes, Abeloos has emphasized, are functions of the absolute size of the planarian and hence are features of *growth, not senescence*. Symptoms of senility are generally results, rather than causes of ageing, and it is in the patterns which lead to the cessation of growth that Abeloos has advised us to seek the causes of the symptoms.

Thus the hypothesis that ageing in the planarian is the result of all those processes of normal (or abnormal) growth which lead to a decline in growth, metabolic rate, and rate of tissue renewal; and that the symptoms of senility are the results of these processes, is consistent with the existing data.

---

* The *Dugesia lugubris—polychroa* complex has been classified karyologically by Benazzi (1957). Biotype B is the tripo-hexaploid member of the diploid to tetraploid (A–D) series of *D. lugubris*. Biotypes E, F, G are reproductively and karyologically isolated from the A–D series and from each other, but are still classified as *D. lugubris* (*in sensu lato*).

† According to Lindh, *D. polychroa*, but in view of the findings of Funaioli (1951), and the karyogram published by Lindh (1959), this material is the *D. lugubris* biotype E of Benazzi (1957). The *D. lugubris* clone 1355 of Lindh is biotype F.

MATERIAL AND METHODS

Material

The animals on which the following study is based were the diploid (A), triploid (B) and tetraploid (D) biotypes (Benazzi, 1957) of *D. lugubris* (O. Schmidt.) Only animals hatched in the laboratory were used so that planarian length could be correlated with age (Reynoldson *et al.*, 1965).

Methods

All material and data were processed as described in an earlier publication (Lange, 1967).

RESULTS

Total number of neoblasts in a planarian as a function of its length. As reported previously (Lange, 1967), the numbers of neoblasts in diploid and triploid planarians were found to vary as an allometric function of the animal's length. The same relationship was found to hold for the tetraploid animals as well. The regression equations were found to be:

$\log N^* = 1{\cdot}90 \log L + 3{\cdot}53$, $\sigma_6 = 0{\cdot}08$, for the diploids†;
$\log N^* = 1{\cdot}93 \log L + 3{\cdot}41$, $\sigma_{11} = 0{\cdot}09$, for the triploids†;
and $\log N^* = 1{\cdot}71 \log L + 3{\cdot}74$, $\sigma_8 = 0{\cdot}03$, for the tetraploids†;

where N^* is the total number of neoblasts per animal (raw count) and L is the length, in mm, of the living animal in the relaxed state.

Analysis of variance for linearity of regression showed that each of the three regressions is valid at better than the 99·9 per cent fiducial level. The slope and intercept of the tetraploid curve were found to be not significantly different from the corresponding values for diploids or triploids, which in turn are not significantly different from each other.

Combining the data for diploid, triploid and tetraploid animals yielded a regression equation of:

$\log N^* = 1{\cdot}95 \log L + 3{\cdot}46$,† $\sigma_{26} = 8{\cdot}46 \times 10^{-2}$, $F_{26} = 1322$.

Cephalo-caudal distribution of neoblasts. Lange (1967) has shown that both the relative position of the pharynx and the cephalo-caudal neoblast distribution in triploid *D. lugubris* remains essentially unchanged as the animal grows from 2 mm in length (size at hatching) to 12 mm in length (adult). This observation implies uniform growth along the cephalo-caudal axis. The pharynx of the diploid, however, was found to shift cephalad with a resultant increase in the relative number of neoblasts in the postpharyngeal region. This implies a non-uniform pattern of growth.

The pharynx positions and neoblast distributions of 5, 8, 10 and 12 mm long tetraploid *D. lugubris* are shown in Fig. 1, and can be seen to be very similar in pattern to that of the triploid.

---

† The values of N^* obtained as solutions to the above equations for given planarian lengths, are overestimates of the true number of neoblasts and must be multiplied by 0·59 to obtain the actual numbers. (q.v. Lange, 1967 appendix.) The value σ, given for each of the equations is the standard deviation of the data about the regresssion line, and the subscript is its number of degrees of freedom.

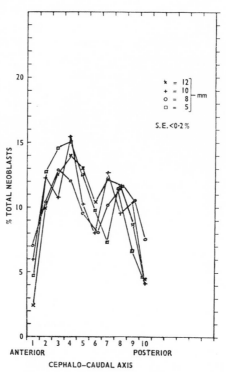

Fig. 1. Percentage of a planarian's (*D. lugubris*, tetraploid) neoblasts to be found in each 10 per cent interval in length along the cephalo-caudal axis. The anterior and posterior limits of the pharynx are found in the middle of the 5th and at the end of the 7th intervals, respectively.

The volume of a planarian (fixed state) as a function of its length. It has already been shown that the volume of triploid *D. lugubris* increases as an allometric function of live length (Lange, 1967).† Further measurements, made as described previously, allowed us to determine the regression equations for diploid and tetraploid animals. The regression equations were found to be:

$$\log V = 2\cdot02 \log L + 7\cdot97, \ \sigma_6 = 0\cdot10, \text{ for diploids,}$$
$$\log V = 2\cdot63 \log L + 7\cdot33\dagger, \ \sigma_{11} = 0\cdot11, \text{ for triploids,}$$
and $\log V = 2\cdot85 \log L + 7\cdot42, \ \sigma_8 = 0\cdot10,$ for tetraploids,

where V is the parenchymal + gut volume in μ^3 (of the fixed animal) and L is the length, in mm, of the living animal in the relaxed, extended state.

Analysis of variance for linearity of regression yielded F values of $F_5 = 258$, $F_{10} = 767$, and $Fc = 175$ for diploids, triploids, and tetraploids respectively, indicating the

† Due to a calibration error, this appeared as $\log V = 2\cdot62 \log L + 5\cdot70$ in our earlier paper, but should have been as given in the text of this paper.

validity of the diploid and triploid regressions at better than the 95 per cent fiducial level. Although the tetraploid regression is not significant, no significant difference in slope or intercept was found between ploidies. The combined data yielded a regression equation of:

$$\log V = 2 \cdot 58 \log L + 7 \cdot 53, \; \sigma_{26} = 0 \cdot 19, \; F_{26} = 44 \cdot 1, \; P < 0 \cdot 05.$$

The slope of the neoblast regression line was found to be very highly significantly different ($t_{52} = 11 \cdot 9$) from that of the volume regression line. Thus, although the number of neoblasts increases with planarian size (age), the parenchymal density of neoblasts decreases.

Planarian length as a function of chronological age. The growth rate and size of a planarian are known to be functions of its past nutrition and environmental temperature (Abeloos, 1930; Reynoldson *et al.*, 1965). The animals used to obtain the preceding data were fed weekly on washed fresh rat liver, *ad libitum*, and kept at a temperature of $20° \pm 0 \cdot 2°C$. Reynoldson (1961) and Reynoldson *et al.* (1965), have shown that *D. lugubris* B fed weekly and kept at $20° \pm 1°C$ produce young which are $2 \cdot 52 \pm 0 \cdot 35$ mm long on hatching and when maintained under the same conditions, grow at a rate (averaged over the time to reach maturity) of $3 \cdot 36$ mm/4 weeks.* As the conditions of our animals were almost identical with those of Reynoldson *et al.*, we can find the

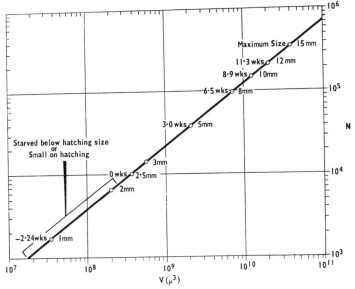

Fig. 2.—Number of Neoblasts (N) [Abercrombie corrected] in a planarian as a function of total fixed tissue volume (V). The live length and age are given for several points.

---

* Reynoldson *et al.* give the figure $4 \cdot 1$ mm/4 weeks, but this value is not consistent with the figures they give for reciprocal maturation time and length at maturity. As the growth rates at other temperatures and for other species are consistent with the corresponding values, the value $3 \cdot 36$ has been derived from those values. Reynoldson (1967) has confirmed that the published value was in error and that the value used here is correct.

equivalent chronological age corresponding to any given planarian size (length). The planarian age in weeks can be estimated to be approximately equal to:

$$4(L - 2\cdot5)/3\cdot36.$$

The relationship between age, planarial tissue volume, and neoblast number. The above regressions of neoblast number and tissue volume on planarial length, allow us to estimate the relationship between tissue volume and neoblast number, and to obtain relative density values. The variation of these parameters as functions of age can be examined by use of the relationship between planarial length and age.

In Fig. 2, we see the variation in neoblast numbers with increasing tissue volume. It must be noted that the volumes in this relationship are those of fixed animals. Lange (1967) has shown that the degree of tissue shrinkage (due to fixation) along the cephalo-caudal axis is proportionally constant (51 per cent $\pm$ 1·5 per cent). Thus live tissue volumes may be as much as 8 (2^3) times those given here for fixed tissues. Until the shrinkage coefficients along the lateral and dorso-ventral axes are known, it is best to consider the relative, rather than absolute, volumes. A time scale has been drawn on the V:N correlation line of Fig. 2. Graphs of tissue neoblast density as functions of length and age are presented as Figs. 3 and 4, respectively.

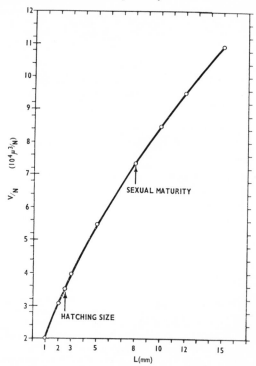

FIG. 3. Rate of decrease of neoblast density (increase of total fixed tissue volume per neoblast) as a function of planarian live length.

106

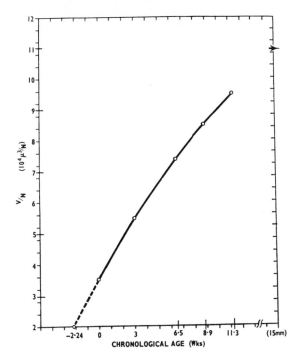

F<small>IG</small>. 4. Rate of decrease of neoblast density (increase of total fixed tissue volume per neoblast) as a function of planarian chronological age (under optimal growth conditions).

CONCLUSIONS AND DISCUSSION

Our main experimental conclusion is that the tissue density of neoblasts (stem cells) decreases as the animal grows larger (and older). The full-grown 10–14 mm animals are by no means old from the point of view of relative completion of lifespan, but they have ceased to grow and we have seen from the literature that their regenerative abilities are less than maximal. We suggest that the decreased regenerative abilities of large adult animals are a result of the observed decrease, relative to total tissue volume, in stem cells —the material from which all regeneration and tissue renewal must come. We know that the adult planarian cannot permanently remain at its full adult size without succumbing to senescence.

Some asexually reproducing species show no signs of ageing, even after 30–40 generations (Lange, unpublished data), and specimens of the normally sexually reproducing *D. lugubris* A and F have been clonally propagated by repeated cuttings and regenerations for periods of many years (Melander, 1963). As starvation and, to a greater extent, regeneration have been shown to have rejuvenatory effects, it would seem that this is accomplished in part by a temporary reduction of tissue volume, accompanied by an increase in neoblast density;* and in part by a decrease in the proportion of old: new cells.

Planarians exposed to LD_{50} doses of X-radiation usually succumb (the 50 per cent that die) within 30–60 days. X-ray doses sufficient to sterilize all neoblasts do not appreciably alter survival time (Lange, in preparation). Thus, 30–60 days could be considered as the lifespan of a planarian (under defined conditions) in which no neoblasts are available for tissue repair and renewal. Neoblasts grafted into sterilized hosts are capable of repopulating the host tissues, but what they will differentiate into is controlled by the host tissues (Stéphan-Dubois and Kolmayer, 1959). In view of the rate of tissue renewal of some vital organ implied by the postirradiation survival time, and as differentiation is controlled by the differentiated tissues of the host, it would appear that differentiation errors, except as cumulative errors, are of much less importance to planarian ageing than the availability of neoblasts. It may be that the shortage of stem cells leads to worn cells not being replaced sufficiently rapidly, which in turn results in improper differentiation control and the symptoms of senility.

ADDENDUM

Benazzi (1967) has recorded the longest observed individual planarian lifespan. This was 21 years (1944–1965) for some individuals of *D. benazzii* (Lepori). However, very few reached this age and, after seventeen years, most became sterile or returned to their original fissiparous state.

Acknowledgements—The author wishes to acknowledge the skilled technical assistance of Miss Judith Whelan and the gift of the original breeding stocks of diploid and tetraploid material from Professor Benazzi of the University of Pisa.

REFERENCES

ABELOOS M. (1930) *Bull. biol. Fr. Belg.* **64**, 1.
BALÁZS A. (1962) *Z. Alternsforsch.* **16**, 53.
BALÁZS A. and BURG M. (1962a) *Acta biol. Hung.* **12**, 287.
BALÁZS A. and BURG M. (1962b) *Acta biol. Hung.* **12**, 297.
BENAZZI M. (1957) *Caryologia* **10**, 276.
BENAZZI M. (1967) Personal communication.
BEST J. B. and RUBENSTEIN I. (1962) *J. comp. physiol. Psychol.* **55**, 560.
BRØNDSTED H. V. (1955) *Biol. Rev.* **30**, 65.
BRØNDSTED H. V. (1956) *Biol. Meddr.* **23**, 1.
CHILD C. M. (1915) *Individuality in Organisms*. Chicago University Press, Chicago.
CURTIS W. C. (1928) *Science* **67**, 141.
CURTIS W. C. and SCHULTZE L. M. (1934) *J. Morph.* **55**, 477.
DUBOIS F. (1949) *Bull. biol. Fr. Belg.* **83**, 213.
ERNHART E. N. and SHERRICK C., JR. (1959) Paper read at Midwestern Psychological Assoc., St. Louis, U.S.A. Cited by JACOBSON A. L. (1963) *Psychol. Bull.* **60**, 74.
FEDECKA-BRUNER B. (1961) *Archs. Anat. microsk. Morph. exp.* **50**, 221.
FEDECKA-BRUNER B. (1964) *C. r. hebd. Séanc. Acad. Sci., Paris* **258**, 3353.
GHIRARDELLI, E. and TASSELLI, T. (1956) *Atti Acad. Sci. Ist. Bologna Memorie Ser. XI*, Tomo **III**, 1.
GOLDSMITH E. D. (1939) *Anat. Rec.* **75**, Suppl. 158.
GOLDSMITH E. D. (1941) *Anat. Rec.* **81**, Suppl. 108.

---

* Lender and Gabriel (1961) state that the neoblasts do not decrease in number in starved planarians. Shrinkage is at the expense of the sexual apparatus, fixed parenchymal elements, and the gastroderm.

HARANGHY L. and BALÁZS A. (1964), *Exp. Geront.* **1**, 77.
HYMAN L. H. (1919) *Biol. Bull. mar. biol. lab. Woods Hole* **37**, 388.
HYMAN L. H. (1951) *The Invertebrates: Platyhelminthes and Rhynchocoela*, Vol. 2, p. 156. McGraw Hill, New York.
JENKINS M. M. and BROWN H. P. (1963) *Trans. Amer. micr. Soc.* **82**, 167.
KOLMAYER S. and STÉPHAN-DUBOIS F. (1960) *J. Embryol. exp. Morph.* **8**, 376.
LANGE C. S. (1966) *J. Embryol. exp. Morph.* **15**, 125.
LANGE C. S. (1967) *J. Embryol. exp. Morph.* **18**, 199.
LE MOIGNE A. (1963) *Bull. Soc. zool. Fr.* **88**, 403.
LE MOIGNE A. (1965*a*) *Bull. Soc. zool. Fr.* **90**, 355.
LE MOIGNE A. (1965*b*) *C. r. Séanc. Soc. Biol.* **159**, 54.
LE MOIGNE A. (1965*c*) *C. r. hebd. Séanc. Acad. Sci., Paris* **260**, 4627.
LE MOIGNE A. (1966) *J. Embryol. exp. Morph.* **15**, 39.
LENDER TH. (1950) *C. r. Séanc. Soc. Biol.* **144**, 1407.
LENDER TH. (1952) *Bull. biol. Fr. Belg.* **86**, 140.
LENDER TH. (1956*a*) *J. Embryol. exp. Morph.* **4**, 196.
LENDER TH. (1956*b*) *Bull. Soc. zool. Fr.* **81**, 192.
LENDER TH. (1956*c*) *Année biol.* **32**, 457.
LENDER TH. (1960) *J. Embryol. exp. Morph.* **8**, 291.
LENDER TH. and GABRIEL A. (1961) *Bull. Soc., zool. Fr.* **86**, 67.
LINDH N. O. (1957*a*) *Ark. Zool.* **10**, 497.
LINDH N. O. (1957*b*) *Ark. Zool.* **11**, 89.
LINDH N. O. (1959) *Ark. Zool.* **12**, 183.
MELANDER Y. (1963) Personal communication.
MCCONNELL J. V., JACOBSON A. L. and KIMBLE D. P. (1959) *J. comp. physiol. Psychol.* **52**, 1.
PALLAS P. S. (1774) *Spicilegia Zoologica quibus novae imprimis et obscurae animalium species iconibus, descriptionibus atque commentariis illustrantur*, p. 22. Fasc. X, Berolini.
REYNOLDSON T. B. (1961) *Oikos* **12**, 111.
REYNOLDSON T. B. (1967) Personal communication.
REYNOLDSON T. B., YOUNG J. O. and TAYLOR M. C. (1965) *J. Anim. Ecol.* **34**, 23.
SCHAROV I. I. (1934) *Trudý Lab. éksp. Zool. Morf. Zhivot.* **3**, 141.
SENGEL Ph. (1951*a*) *Bull. biol. Fr. Belg.* **85**, 376.
SENGEL Ph. (1951*b*) *C. r. Séanc. Soc. Biol.* **145**, 1381.
SENGEL Ph. (1953) *Archs. Anat. microsc.* **42**, 57.
STÉPHAN F. (1962) *C. r. Séanc. Soc. Biol.* **156**, 920.
STÉPHAN-DUBOIS F. and KOLMAYER S. (1959) *C. r. Séanc. Soc. Biol.* **153**, 1856.
VOIGT W. (1928) *Zool. Jb.* **45**, 293.

Summary—The literature on planarian ageing is briefly reviewed to distinguish those features which are symptoms of senility from those which are features of normal growth and development. The recommendation of Abeloos, to seek the causes of ageing in those processes of normal development which lead to cessation of growth, has been followed and the relationships between planarian chronological age and size, number of stem cells (neoblasts) and tissue density of stem cells have been examined. Quantitative data are presented on these parameters for the species *Dugesia lugubris* (O. Schmidt).

The relevance of the drastic decrease in tissue neoblast density as the planarian increases in size is discussed in terms of the role of the neoblast in normal tissue renewal and in regeneration. It is concluded that although cumulative differentiation errors may contribute to the symptoms of senility, neoblast availability for regeneration and repair is probably the critical factor in planarian survival and that a shortage of neoblasts may even be the cause of the improper differentiation control by the differentiated tissues which results in the symptoms of senility.

Studies on the cellular basis of radiation lethality
I. The pattern of mortality in the whole-body irradiated planarian (Tricladida, Paludicola)

C. S. LANGE

The planarian lends itself to parallel cellular and whole animal studies because it contains a single, morphologically identifiable population of toti-potential stem cells, without which the animal cannot long survive.

Two genera, including five species and three auto- or allopolyploid biotypes, of planarian have been chosen for a study of the cellular basis of radiation lethality. The distribution of mortality times was found to be uni-modal and the x-ray dose-response curve, sigmoid. By the use of antibiotics, infection by commensal organisms was shown not to be a significant factor in mortality. The effects of several other treatments are interpreted in terms of altered metabolic rate. It is suggested that planarian radiation mortality is the end-result of a loss of reproductively-intact stem cells leading to insufficient replacement of worn parts of vital tissues.

1. Introduction

Radiobiology has, for technical reasons, been divided into two fields of study. In one the object studied is the *cell* (*in vitro* and/or *in vivo*), in the other, the *organism*. In the planarian it is possible to study both cell and organism, because the planarian contains a single, morphologically identifiable population of toti-potential stem cells, without which the organism cannot long survive. None of the differentiated cells of the planarian is known to be capable of mitosis. More-over, this stem-cell population, although clonal in nature, grows subject to the mechanisms of normal growth control, i.e. on demand. It is also subject to the activity of differentiator substances and, by differentiation, provides the cells for all tissue replacement. Because the planarian consists entirely of soft tissue, the problem of bone dosimetry does not arise. The existence of auto or allo-polyploidy in the Planariidae is yet another advantage in using them as material for radiobiological studies.

In this series of papers the problems of metabolic rate, infection, regeneration, age, oxygenation and polyploidy as they affect radiation lethality will be examined. The survival probability of the organism will be related quantitatively to the survival of reproductive integrity of the stem cell.

2. Materials and methods
2.1. *Materials*

The planarians used are small (~ 1 cm long) freshwater animals noted for their high regenerative capacities. The species studied include the diploid, triploid and tetraploid biotypes of *Dugesia lugubris* (O. Schmidt) (respectively, biotypes A, B and D according to Bennazzi 1957); diploid and aneuploid popu-lations of *D. etrusca* (Benazzi); *D. tigrina* (Girard); *Polycelis tenuis* (Iijima) (the *Polycelis tenuis-hepta* complex of Hansen-Melander, Melander and Reynoldson 1954, which Benazzi 1963, demonstrates to be a single species with several incipient species *in statu nascendi*); and *Polycelis felina*.

Of the auto- or allopolyploid *D. lugubris* series, only the triploid biotype B is commonly found in the British Isles and northern Europe. Animals of this biotype were collected in Shropshire. The diploids, biotype A, are to be found only in Italy (Benazzi 1957) and Sweden (Melander 1963). The specimens of diploid *D. lugubris* used in this study were collected from an artificial pond in the Botanical Garden of the Department of Botany of the University of Pisa.

The tetraploid material, which is extremely rare in nature, was grown from the descendants of two parthenogenetic tetraploids produced by Professor Benazzi and is shown by the crosses in figure 1 (Benazzi, personal communication 1964). This species is oviparous.

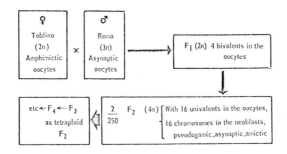

Figure 1. Origin of the tetraploid *D. lugubris* biotype (after Benazzi 1964).

Both populations of *D. etrusca* are native only to Italy and were kindly provided by Professor Benazzi. This species has both oviparous and fissiparous stages in its life-cycle, but only fissiparous reproduction was allowed in our populations. Only fissiparous members of this species were used for experiments.

Populations of *D. tigrina* and *Polycelis tenuis* were collected in Shropshire, and populations of *Polycelis felina* from a spring near Hastings (Ord. Surv. TQ7916). *D. tigrina* and *Polycelis felina* were kept in the fissiparous form; *Polycelis tenuis* is naturally oviparous.

Chromosome analysis was used routinely to confirm the purity of each population of the *Dugesia lugubris* biotypes.

2.2. Culture methods

2.2.1. Containers

Three types of container were used. Animals which had to be kept isolated, either for experimental reasons, or to prevent sexual reproduction between members of a clone, were placed in disposable 13 ml. or 30 ml. crystal polystyrene tubes (Metal Box Co.) with perforated polythene caps. All other animals were kept in glass bottles (squat type, Solmedia, Ltd.) at ≈ 30–50 adults per 400 ml. bottle.

2.2.2. *Water*

All animals were kept in water brought into the laboratory directly from the mains via iron and lead plumbing. This is effectively spring water and will be called 'direct mains water', remembering that copper or brass pipes may cause cupric poisoning. The concentration of chlorine in Manchester water (0·1–0·4 p.p.m. to the consumer) is not toxic to planarians.

2.2.3. *Food*

Animals were fed once or twice a week on rat or beef liver, cut into small pieces, the water being changed and waste products removed the following day. Tubes and bottles were replaced as soon as they showed signs of becoming dirty. Occasionally animals were fed with tubifex worms (*Tubifex tubifex*), although this was done mainly for freshly hatched young. Tubifex is a particularly good food for experimental animals, since it is possible to observe which animals ate and how much.

2.2.4. *Temperature*

All animals were kept in a 20 °c incubator, the air temperature of which varied by less than ± 1°c. Temperature fluctuations of the water in the incubator were much less than that ($< \pm 0.2$°c).

2.2.5. *Handling*

All manipulation of the animals was done by means of a pipette or an artist's paintbrush.

2.2.6. *Population growth-rates*

The population growth-rates of the fissiparous *D. etrusca* are shown in figure 2. Tubifex and liver are seen to be equally good foods, whereas brain was obviously deleterious. The rate of increase for the pseudogamic tetraploid *D. lugubris* D is shown in figure 3. The rates of increase of the cocoon-laying species are deceptively high, in that after hatching, there can be a high mortality rate until animals reach adult size. Amongst the triploid *D. lugubris* B, this is mainly due to cannibalism.

2.3. *Irradiation methods*

All x-ray irradiations were carried out on a Resomax 300 machine. The physical parameters were: 300 kVp; 19·5 mA; filtration 1·0 mm Al, h.v.l. 1·9 mm Cu. Crystal polystyrene irradiation dishes were used to restrain the planarians to a small volume of water surrounded by water-equivalent scattering material.

In view of the report of Paterson (1942) to the effect that irradiated Perspex (Polymethyl methacrylate) and Distrene (an I.C.I. product) give off substances toxic to cells grown *in vitro*, the toxicity of irradiated polystyrene was tested on both HeLa cells and planarians. No toxicity was detected after several hours exposure (longer than any irradiation time used) to previously heavily irradiated (10 krads in water) water-filled crystal polystyrene containers.

Dosimetry was carried out with a Baldwin–Farmer ionization chamber and with LiF. The probable error for beam flatness was ± 1·3 per cent, for field depth (± 2·5 mm water) was ± 1·3 per cent, and for x-ray output was ± 1 per cent, giving a total probable error of ± 2·1 per cent.

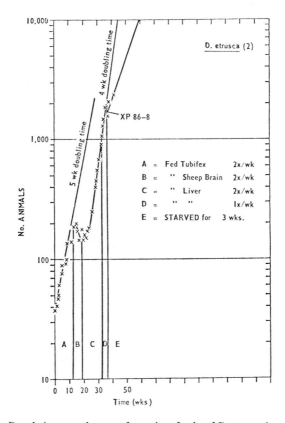

Figure 2.   Population growth curve, for various foods, of *D. etrusca* (population 2).

Animals to be irradiated were placed, up to 20 or 30 per dish, in the irradiation dishes and irradiated as shown in figure 4. The dose-rate to the top dish was equal to 195 soft tissue rads/min. The difference in x-ray quality between the top and bottom dishes was of negligible importance with respect to the biological response observed. After irradiation, unless otherwise stated, the animals were removed from the irradiation dishes and placed in individual 13 ml. crystal polystyrene tubes filled with direct mains water. The tubes were then capped with perforated polythene caps and placed in the 20°c incubator. In the experiments in which the head-frequency (percentage of head regeneration) was determined, the specimens were examined daily; for LD_{50} determination they were examined thrice weekly for the duration of the experiment—usually 100–150 days after irradiation.

113

2.4. *Methods of data analysis*

2.4.1. *Criterion for survival*

All the species used in the experiments reported below, possess well-developed powers of regeneration: even a small piece of a planarian is capable of becoming a complete animal and growing to adult size again. Thus, if even as little as 1 per cent of the animal by volume remained intact (unlysed), it would still have survived the irradiation. As a demonstration of this, in a few cases such small pieces were allowed to grow to adulthood.

Thus the criterion for mortality used in all experiments was that the entire animal be lysed with no piece left intact. Until the last piece had lysed, the animal was considered a survivor.

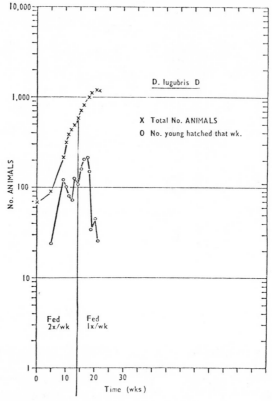

Figure 3. Population growth curve of *D. lugubris* D (tetraploid).

2.4.2. *Estimation of animal survival curve parameters*

For the endpoint time chosen (see results below) the planarian x-ray dose–survival curve is sigmoid and can therefore be fitted in probits by the Maximum Likelihood Method of Finney (1962). This procedure was computerized and the equation fitted was of the form: $y = mx + b$, i.e. probit $(P - C/1 - C) = (1/K)$ (dose) $+ (- LD_{50}/K)$ where P is the observed mortality probability, C is the natural mortality probability, K is the inverse slope of the probit line, and LD_{50} is the intercept of the probit line at 50 per cent mortality. The argument of the probit function, $(P - C/1 - C)$, is known as the Abbot kill, or Abbott's correction for natural mortality.

The input data consist of the doses given (including the control dose of zero), the number of animals at risk for each dose, and the number of animals affected (mortality) at each dose. The output consists of the two parameters of the probit line, the LD_{50} and K, and the natural mortality, C, together with their respective standard errors, the total χ-square and its degrees of freedom, the number of iterations performed, and the variance : covariance matrix for the fit. A minimum of three iterations were carried out, and more if necessary to reduce the parameter changes from the previous cycle to less than one standard error (S.E.) of each parameter, up to a maximum of nine iterations.

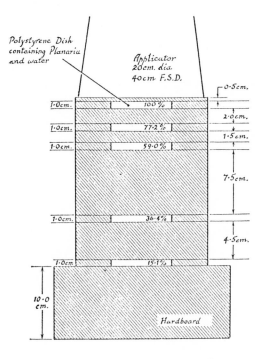

Figure 4. The set-up for x-irradiations.

The standard errors of fit may be slight overestimates, as they include the correction factor χ-square per degree of freedom to allow for the case where significant, but not systematic, deviations of the data from the fitted line exist. This is discussed by Finney (1962). To prevent underestimation of the standard errors by this correction factor, for those cases where the χ-square per degree of freedom falls below unity, this ratio was arbitrarily set at unity, as lower values would be fortuitous in data expected to fit a Gaussian distribution. For a more detailed description of the computer programme, see Gilbert (1968).

3. Results

3.1. *The time-course of post-irradiation mortality and the selection of endpoints*

The time-course of post-irradiation mortality for a typical x-ray survival experiment is shown in figure 5. In the dose range $LD_{50} \pm LD_{50}$, the graph of the proportion of animals surviving (ordinate) as a function of time after irradiation (abscissa) follows the same pattern. For the species (*D. etrusca*) used in this experiment, there is no change in the proportion of survivors (although necrotic processes may appear) during the first 30 days after irradiation. Death (complete lysis) occurs between day 30 and day 45 (period of mortality). After day 45, the proportion surviving remains constant. Animals were routinely examined for 100–150 days to confirm the constancy of the proportion surviving after day 45. In some cases, survivors were watched for nearly a year before being discarded. Survival for 60 days was therefore taken as a general endpoint time. Deviation from this pattern was found to occur only in the species *Polycelis tenuis*, in which the period of mortality extended over a considerably longer period. Furthermore, the exact time of death for each animal was difficult to establish, as this species encapsulates (figure 6) in a thick sheath of mucus. Recovery from the radio-lesions occurs, or fails to occur, while the animal is encapsulated, and lysis is slow and progressive.

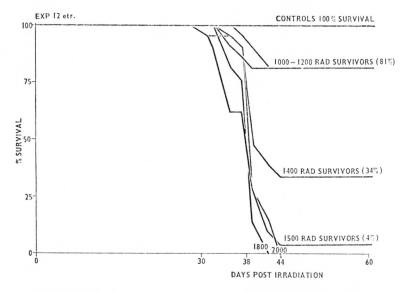

Figure 5. The time course of post-irradiation mortality—from a typical experiment.

The choice of a 100 or 120 day endpoint time for *Polycelis tenuis* removes the anomalies due to artefacts and yields the normal pattern of the dose–response curve.

Because of the narrowness of the uni-modal mortality time distribution, the difference between the geometric mean time of death (log time versus mortality distribution) and the arithmetic mean time of death, was not significant. Therefore, for convenience, the arithmetic mean was used.

116

The parameters of survival, LD_{50}, K and t_m (mean survival time) were found to be constants for any given set of experimental conditions, i.e. the results of successively repeated experiments were collectively in agreement.

3.2. The dose-dependence of post-irradiation mortality

For any endpoint time well past the modal time of death, the planarian x-ray dose–survival curve was found to be sigmoid.

Although the sigmoid patterns of both the time-course and the dose dependence of post-irradiation mortality were a constant finding, the mean time of death (t_m), and the mean lethal dose (LD_{50}) were not invariant to environmental changes such as nutrition.

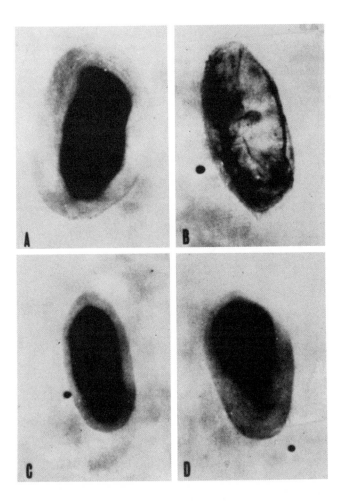

Figure 6. Mucus capsules of *Polycelis tenuis*. (A) and (C) planarian in capsule, (B) empty capsule after recovered planarian has left, (D) capsule with slowly lysing planarian in it.

117

Planarians in their natural environment are subject to a variable food supply and can tolerate starvation for up to 4 or 5 months with no adverse effects other than shrinkage due to the catabolism of their own differentiated tissues (Rosenbaum and Rolon 1960, Lender and Gabriel 1961). As a planarian's metabolic state is a function of its food supply and the temperature of its environment (Hyman 1951, Brøndsted 1953), experiments to determine the effect on survival of feeding before and after irradiation were performed.

3.3. The effect of feeding after irradiation

The contrast in the effects of feeding and starvation after irradiation is most marked in the species *D. etrusca* (see tables 1 and 2, and figure 7). After 2000 rads (in soft tissue), a dose at which the proportion surviving is well under 5 per cent, fed animals showed a 30–40 per cent decrease in survival time compared with starved animals (table 1). For all doses in the $LD_{50} \pm LD_{50}$ dose range, time of death decreased from 43 or 39 days (median and arithmetic mean respectively) in

	50 per cent dead	100 per cent dead	Percentage of survival 60 days
Fed (2000 rads)	22 days	32 days	0
Starved (2000 rads)	38 days	48 days	0
Fed (1000 rads)	22 days	—	2·5
Starved (1000 rads)	—	—	72·5

Table 1.   Effect of post-irradiation feeding on survival of *D. etrusca*.

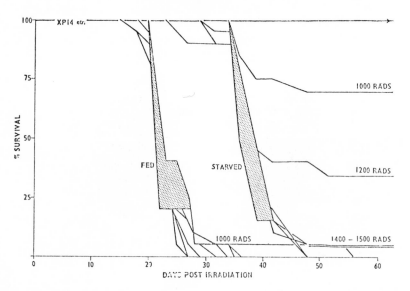

Figure 7.   Effect of feeding after irradiation on the survival of *D. etrusca*—from a typical experiment.

118

Species	Condition	Number of experiments	f_t*	$t_m \pm$ S.E.† (days)	LD$_{50}$ ± S.E. (rads)	$K \pm$ S.E. (rads)	f_s‡	$C \pm$ S.E. (per cent)
D. etrusca	Starved	4	47	39·1 ± 1·0	1207 ± 30	269 ± 27	28	1·1 ± 1·2
	Fed	2	38	25·1 ± 1·2	§757 ± 854	124 ± 433	11	0 ± 1·6
D. lugubris B	Starved	7	52	27·4 ± 0·7	1677 ± 143	936 ± 206	49	12·0 ± 5·5
	Fed	2	9	22·5 ± 1·3	1502 ± 96	365 ± 105	8	8·1 ± 4·3
Polycelis tenuis	Starved	5	37	58·4 ± 4·9	‖706 ± 104	388 ± 94	36	1·9 ± 3·1
	Fed	2	39	30·6 ± 2·0	¶718 ± 40	272 ± 49	9	2·5 ± 2·7

* Degrees of freedom of t_m.
† At LD$_{50}$ dose.
‡ Degrees of freedom of LD$_{50}$, K, C.
§ The large standard errors for the estimates of the probit line parameters for the fed population are due to an insufficiency of data at low doses. The means, however, serve as useful guides to the magnitude of the effect, confirmed by the significance of the difference in survival level ($P < 0.001$) at 1000 rads (see table 1).
‖ At 120 days after irradiation.
¶ At 60–120 days after irradiation.

Table 2. Survival-curve parameters and survival-time comparisons for animals fed versus starved after-irradiation.

the starved animals to 27 or 25 days. This difference is highly significant ($P < 0.001$). After 1000 rads (soft tissue) the proportion surviving was also visibly affected, falling from 74 per cent (starved) to 2·5 per cent (fed). This difference was also highly significant ($P < 0.001$). Thus feeding after irradiation is extremely deleterious to the survival of *D. etrusca*.

D. lugubris B shows a small but real decrease in survival time if fed after irradiation ($0.05 > P > 0.01$). The mean time of death (after doses of the order of the LD_{50}) drops from 27.4 ± 0.7 days to 22.5 ± 1.3 days after irradiation. The slopes of the survival curves also differ significantly ($0.05 > P > 0.01$), becoming steeper after feeding (table 2). The LD_{50}, however, does not decrease significantly. Thus feeding after irradiation can be said to be slightly deleterious to the survival of *D. lugubris* B, in that the survival time was decreased and that at doses somewhat greater than the LD_{50}, the proportion surviving also decreased.

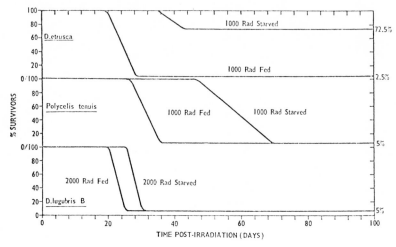

Figure 8. Effect of feeding after irradiation on survival time and level.

The survival curves for *Polycelis tenuis* show no significant changes in LD_{50} or slope after feeding, if reasonable endpoint times are chosen for the two groups (60 days if fed, 120 days if starved); but the decrease in survival time, from 58.4 ± 4.9 days to 30.6 ± 2.0 days, is highly significant ($P < 0.001$). Thus the only effect of feeding after irradiation on *Polycelis tenuis* is to alter the rate of mortality with time without affecting the absolute level of survival at any dose. An additional effect of feeding is that encapsulation is markedly reduced, thus simplifying mortality estimation.

Figure 8 shows three graphs which describe the effects of x-ray doses in the $LD_{50} \pm LD_{50}$ region on fed and on starved animals of each of the three species studied. For *D. lugubris*, feeding makes only a small change in survival time; for *Polycelis tenuis*, survival time is considerably altered, but no change in the proportion surviving is found; while for *D. etrusca*, changes in both survival time and the proportion surviving are found. The possible relation of these species differences to basal metabolic rates is discussed below.

R.B.   2 0

3.4. *Starvation before irradiation*

The possibility that starving before irradiation might have a deleterious effect was suggested by the results of some early experiments with *D. etrusca*. The LD_{50} for a population of animals irradiated after a 20 day period of starvation was significantly lower than that for one irradiated after a 16-day starvation period ($0.05 > P > 0.01$). Even shorter starvation periods appeared to produce a greater heterogeneity of response to the irradiation, as might be expected if the duration of the starvation period was related to survival (LD_{50}). However, further experiments using shorter starvation periods, showed a small but significant oscillation in LD_{50}; and attempts to repeat these observations in a second, non-clonal, population of *D. etrusca* failed to demonstrate any correlation between duration of starvation (up to 20 days) before irradiation and survival (LD_{50}) after irradiation. The LD_{50} oscillation observed in the first population (clonal) and the constancy of LD_{50} observed in the second population (non-clonal) are shown in figure 9. The probit line parameters and mean times of death for each of the starvation periods are given in table 3.

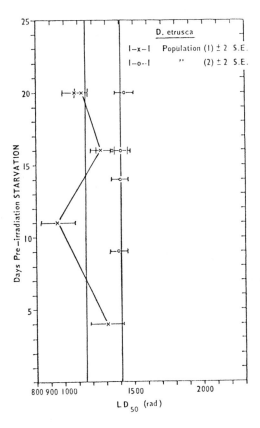

Figure 9. Effect of duration of starvation before irradiation on LD_{50} of *D. etrusca*.

The duration of the pre-irradiation starvation period was found to have no effect on the LD_{50} for any of the other species studied (table 3).

On the basis of the results of the starvation experiments, a routine of one or more weeks starvation before, and continuous starvation after irradiation (until well after the endpoint time), was followed for all the other experiments.

Experiment number	Days prestarved	f_s*	$LD_{50} \pm$ S.E. (rads)	$K \pm$ S.E. (rads)	$t_m \pm$ S.E.† (days)	f_t‡	$C \pm$ S.E. (per cent)
D. etrusca (1) (clonal)							
1001	4	8	1311 ± 60	262 ± 57	$37 \cdot 6 \pm 1 \cdot 6$	4	$0 \pm 1 \cdot 3$
702	11	3	961 ± 61	148 ± 85	$36 \cdot 2 \pm 7 \cdot 6$	4	$2 \cdot 5 \pm 1 \cdot 8$
1200	16	4	1272 ± 36	208 ± 36	$39 \cdot 5 \pm 0 \cdot 5$	17	$0 \cdot 4 \pm 2 \cdot 1$
1401	20	4	1090 ± 44	219 ± 50	$39 \cdot 8 \pm 1 \cdot 4$	19	$0 \pm 2 \cdot 2$
D. etrusca (2) (non-clonal)							
92	9	5	1369 ± 30	185 ± 33	$36 \cdot 9 \pm 1 \cdot 0$	17	$1 \cdot 2 \pm 1 \cdot 3$
12000	9	2	1289 ± 163	57 ± 241	$24 \cdot 0 \pm 0 \cdot 6$	24	$0 \pm 2 \cdot 2$
12403	9	2	651 ± 936	1361 ± 1140	$31 \cdot 6 \pm 1 \cdot 5$	4	$0 \pm 3 \cdot 2$
10701	12	6	1398 ± 29	178 ± 35	$41 \cdot 8 \pm 2 \cdot 9$	12	$0 \pm 2 \cdot 2$
93	14	5	1413 ± 29	159 ± 31	$32 \cdot 7 \pm 0 \cdot 6$	30	$0 \pm 1 \cdot 1$
10801	14	2	1323 ± 28	108 ± 26	$30 \cdot 0 \pm 1 \cdot 8$	23	$0 \pm 2 \cdot 2$
87	16	5	1366 ± 61	220 ± 62	$39 \cdot 0 \pm 0 \cdot 8$	23	$0 \pm 2 \cdot 1$
96	16	5	1419 ± 25	118 ± 23	$31 \cdot 6 \pm 0 \cdot 3$	28	$0 \pm 1 \cdot 1$
99	20	4	1310 ± 32	174 ± 31	$35 \cdot 4 \pm 0 \cdot 6$	19	$0 \pm 1 \cdot 6$
D. lugubris B							
701	5	8	1201 ± 61	150 ± 53	$21 \cdot 4 \pm 1 \cdot 5$	6	$0 \pm 1 \cdot 6$
6301	5	6	1737 ± 211	408 ± 222	$30 \cdot 1 \pm 2 \cdot 1$	9	$22 \cdot 3 \pm 13 \cdot 2$
7601	5	4	1151 ± 497	2285 ± 905	$29 \cdot 6 \pm 0 \cdot 8$	23	$15 \cdot 3 \pm 8 \cdot 5$
3400	6	3	1757 ± 292	972 ± 440	$24 \cdot 1 \pm 2 \cdot 1$	8	$0 \pm 3 \cdot 3$
4703	20	4	1772 ± 491	1184 ± 1360	$30 \cdot 1 \pm 2 \cdot 9$	13	$0 \pm 8 \cdot 1$
5212	20	2	1400 ± 366	400 ± 343	$23 \cdot 8 \pm 1 \cdot 2$	13	$39 \cdot 8 \pm 21 \cdot 1$
3800	29	4	1855 ± 127	264 ± 163	$27 \cdot 9 \pm 1 \cdot 2$	11	$26 \cdot 7 \pm 10 \cdot 5$
D. lugubris A							
7501	5	4	2907 ± 93	356 ± 126	$29 \cdot 2 \pm 1 \cdot 9$	15	$7 \cdot 5 \pm 2 \cdot 8$
7701	5	6	2400 ± 204	760 ± 233	$37 \cdot 4 \pm 1 \cdot 9$	23	$0 \pm 3 \cdot 1$
5112	20	2	2098 ± 307	703 ± 471	$21 \cdot 8 \pm 4 \cdot 1$	4	$2 \cdot 2 \pm 5 \cdot 4$
Polyc lis tenuis							
6201	5	5	981 ± 235	564 ± 257	$47 \cdot 2 \pm 11 \cdot 6$	5	$8 \cdot 6 \pm 12$
4302	20	5	540 ± 256	265 ± 179	$63 \cdot 1 \pm 5 \cdot 3$	10	$0 \pm 3 \cdot 2$
4903	20	4	726 ± 279	353 ± 206	$47 \cdot 9 \pm 9 \cdot 2$	7	$0 \pm 3 \cdot 2$
5304	20	6	800 ± 135	151 ± 150	$58 \cdot 4 \pm 13 \cdot 9$	4	$0 \pm 7 \cdot 1$
5401	20	4	581 ± 433	786 ± 383	$70 \cdot 8 \pm 11 \cdot 2$	7	$0 \pm 4 \cdot 4$

* Degrees of freedom of LD_{50}, K, C.
† At LD_{50} dose.
‡ Degree of freedom of t_m.

Table 3. Survival-parameters and survival-time comparisons for different durations of starvation before irradiation.

To test the hypothesis that death after irradiation might be due, in part, to loss of ability to control commensal flora (an important factor in mammalian mortality—Byron and Lajtha 1966), the experiments which follow were performed.

3.5. *Effects of antibiotics*

Of the various antibiotic treatments designed to control or rid the planarian of its commensal flora, which have been described in the literature, only the following two did not prove lethal to the planarian species used in this study:

(1) Animals were placed in tubes of direct mains water to which chloramphenicol had been added to a final concentration of 25 μg/ml.—a concentration shown by Henderson and Eakin (1959) to have no effect on planarian regeneration, but sufficient to prevent growth of the flora commensal to the planarian (*D. dorotocephala*).

(2) Planarians were washed in two changes of a penicillin–streptomycin solution (2500 units of each/ml.) followed by several sterile distilled-water rinses and were then returned to tubes containing sterilized direct mains water. This, the method of Ansevin and Buchsbaum (1961), was reported to be sufficient to provide aseptic explants for *in vitro* culture.

Both treatments were used on populations of *D. etrusca* in a comparison of the relative effects of the removal of commensal organisms by penicillin + streptomycin, and of the bacteriostasis produced by chloramphenicol. The effect of the antibiotic treatments, when given within a few minutes to hours after the irradiation, was to delay slightly (14–21 per cent) the time of death of those destined to die; but the final level of survival at any one dose was not significantly changed. In both cases, no significant change was found in the parameters of the probit survival curves, but the increase in survival time was found to be highly significant ($P < 0.005$). The penicillin + streptomycin treatment yielded a greater increase in survival time than did the chloramphenicol treatment. This difference was significant ($0.05 > P > 0.01$) (see table 4).

To confirm that the antibiotics used were effective in controlling the organisms commensal to the planarians in our laboratory, identification and sensitivity tests were performed.

(*i*) *Identification of organisms*. Tubes containing (1) direct mains water, (2) direct mains water + food, (3) direct mains water + food + *D. lugubris*, (4) direct mains water + food + *Polycelis felina*, (5) direct mains water + lethally irradiated *D. lugubris*, and (6) a mélange of irradiated planarian lysates, were set up and their contents allowed to lyse. Samples from each tube were cultured in nutrient broth at room temperature, and then plated for identification and sensitivity. The nutrient broth culture consisted of a 1 : 1 mixture of the sample to be tested and a sterile solution of Oxoid nutrient broth No. 2 (CN67) at a concentration of 50 g/ml. tap water. Tubes from groups (1) and (2) were found to contain gram-negative bacilli and gram-positive cocci. Tubes to which planarians had been added were also found to contain *Pseudomonas aeruginosa* and *Streptococcus* α-*haemolyticus*. The mélange of lystates was also found to contain a psychrophilic *Pseudomonas* and some *Klebsiella*.

(*ii*) *Sensitivity to antibiotics*. Of the antibiotics tested. (Oxoid Multodisks 30–9C and U–2; see table 5), only kanamycin and colomycin controlled all of the organisms; kanamycin was effective at a concentration of 1 μg/ml.

Tests of the sensitivity of planarians to various concentrations of kanamycin showed that concentrations of 10 μg/ml. and 100 μg/ml. have no effect on the survival and/or the rate of head regeneration of *D. lugubris* B and *D. etrusca* (exposed for 67 days). For the x-irradiation experiments in which kanamycin

Species		Treatment	LD$_{50}$±S.E. (rads)	K±S.E. (rads)	f_s*	t_m±S.E. (days)	f_i†
D. etrusca	(1)	Control	1090±44	219±50	4	39·8±1·4	19
	(1)	Chloramphenicol	1112±39	203±43	4	45·2±1·0	18
	(1)	Penicillin+streptomycin	1191±38	218±42	4	48·2±0·9	12
	(2)	Control	1359±19	163±18	20	35·1±0·5	72
	(2)	Kanamycin	1408±30	214±32	20	41·2±0·6	79
	(1)	Feeding+chloramphenical ⎤	757±854‡	124±433‡	11	24·4±1·9	18
	(1)	Feeding control ⎦				25·7±1·6	19
D. lugubris B		Control	1677±143	936±206	49	26·2±0·7	97
		Kanamycin	1971±123	511±140	4	47·1±2·7	19

* Degrees of freedom of LD$_{50}$, K, C.

† Degrees of freedom of t_m.

‡ Curve fitting errors are abnormally high as a result of the paucity of survivors of the lowest irradiation dose used (1000 rads).

Table 4. Effects of antibiotics on survival.

124

was used to control commensal organisms, it was used at a concentration of 10 μg/ml.

As was found for the antibiotics previously tried, the parameters of the probit survival curve were not changed significantly, but the survival time was increased by 17 per cent and 81 per cent for *D. etrusca* and *D. lugubris* B, respectively (see table 4). These increases in survival time were found to be highly significant ($P < 0.001$).

Thus planarian death after irradiation is not, to any significant degree, a result of infection rather than loss of reproductively competent stem cells (neoblasts).

Antibiotic*	Concentration	Resistant†
Kanamycin	30 μ/disc	None
Kanamycin	1 μg/ml.	None
Colomycin	200 μg/disc	None
Noxytiolin	10 mg/disc	5
Negram		2, 4
Chloramphenicol	50 μg/disc	4
Chloramphenicol	25 μg/ml.	1, 3–5
Penicillin and streptomycin	2500 units each/ml.	1, 4, 5
Nitrofurantonin	200 μg/disc	3–5
Penbritin		1, 3–5
Neomycin	10 μg/disc	2–5
Streptomycin	25 μg/disc	1, 3–5
Cloxacillin	5 μg/disc	1, 3–5
Sulphonamide		2–5
Nystatin	50 units/disc	1–5
Penicillin	2500 units/ml.	1–5

* Oxoid Multodiscs 30–9C and U–2.
† Source of resistant organisms: 1, water only; 2, water + food; 3, water + food + planarian (lysed); 4, water + irradiated planarian (lysed); 5, mélange of irradiated planarian lysates.

Table 5. Control of commensal organisms by antibiotics.

3.6. *Effect of antibiotics on the post-irradiation feeding effect*

The presence of antibiotics after irradiation did not significantly influence the effects of post-irradiation feeding (see table 4). The parameters of the probit survival curve and the time of death were not changed significantly. Survival at each dose level was not improved by the presence of antibiotics. Thus the effects of feeding after irradiation were not a result of an increased susceptibility to infection by commensal organisms at a time when the planarian's ability to deal with such organisms may have been diminished.

4. Conclusions and discussion

The planarian was chosen for this study because of its anatomical and physiological organization. The time-course of radiation mortality has been shown to be uni-modal, and reproducible for a given set of experimental conditions. This uni-modality may be indicative of a single mortality syndrome, in much the same way that the bi-modality of mammalian radiation-mortality (in the appropriate dose range) is indicative of 'bone-marrow' and 'gut' syndromes. The dose-dependence of radiation mortality is sigmoid and the parameters of the survival curve have highly reproducible values.

Bond, Fliedner and Archambeau (1965) have ably advocated the thesis that mammalian radiation lethality is to be understood basically in terms of the cellular kinetics of maintenance and renewal of the organs and organ-systems essential for life. As a framework against which we shall try to understand the results of this study, this thesis has been extended (see below) to cover all multicellular organisms.

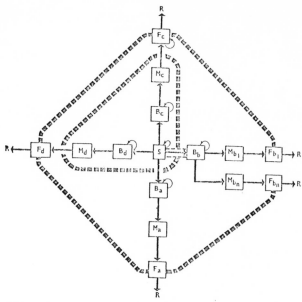

Figure 10. Schematic representation of an organism. The square ' S ' in the centre represents the stem-cell compartment, which at its earliest stages consists of the zygote and then the blastomeres. The arrow leaving the square and returning represents mitotic renewal (maintenance). The arrows leaving the square represent differentiation. The inner dashed ring represents the period of embryogenesis and the outer dashed ring the organism. B = blast cells, M = maturation, F = functional tissues, R = removal of surplus or worn and damaged cells. The subscript 'a' line is an example of post-embryonic tissues deriving from uni-potential blast cells. The ' b ' line represents post-embryonic tissues deriving from pluri-potential blast cells. The dashed corridor linking S to B indicates that in some cases the pluri-potential blast cells may derive from post-embryonic toti-potential stem cells. The ' c ' line represents post-embryonic self maintenance of a differentiated tissue by mitosis (e.g. liver) and the ' d ' line represents non-replicating post-embryonic differentiated tissues (e.g. nerve).

The organism schematically portrayed in figure 10 starts with a stem cell (the zygote), which reproduces itself. In some cases, there will always be a toti-potential population of stem cells, (e.g. Planariidae). In other cases all of the stem cells may differentiate to form pluri or uni-potential blast cells (e.g. blast cells of ectoderm, mesoderm and endoderm; or the erythroblasts, myeloblasts and megakaryoblasts of the bone-marrow). In others, all the blast cells (of a given line) may differentiate to produce self-replicating functional cells (e.g. liver), and in a few cases all the blast cells (of a line) may differentiate to produce non-replicating functional cells (e.g. mammalian nerve). The organism is able to survive so long as sufficient functional cells remain to carry out the vital functions of each of the essential organ systems. Thus the lower the rate of wear and tear (and thus of renewal) and the higher the excess of functional cells above the bare minimum consistent with continued function of such an organ-system, the longer the latent period preceding the expression of damage to precursor cells in the form of organ failure and organismal death. As with the weakest link of a chain, the weakest part of an essential organ or organ-system determines the time of mortality. While the organ-system continues to function, it is the degree of depopulation (in terms of loss of reproductive integrity) of the relevant precursor cells, and the rate at which they repopulate their own compartments (repair) and provide new functional cells (replacement), that will determine the long-term survival of the organism. A low functional cell attrition rate and hence a low differentiation demand in the early post-irradiation period may make possible a more rapid repair and thus a greater future ability to replace worn functional cells. Once repair (repopulation) has occurred, higher attrition rates can be tolerated.

Returning to the results of the experiments presented in this paper, it is noted that variation of the pre-irradiation metabolic rate by feeding or starvation has no effect on time of death nor on the proportion surviving (i.e. LD_{50} and K). However, a decreased *post-irradiation* metabolic rate induced by starvation post-pones the time of death in all species. This might be expected if the metabolic rate determines the rate of attrition (and rate of demand for renewal) of those organ-systems of which the functioning is a prerequisite for survival. Only the fissiparous species (*D. etrusca*) showed an increase in the proportion surviving with decreased post-irradiation metabolic rate.

The difference in response observed following treatments which alter the post-irradiation metabolic rate for oviparous versus fissiparous planarians need not imply different metabolic patterns. The same result might be expected to follow from a quantitative difference of sufficient magnitude to yield a qualitatively different result. Thus the fissiparous planarians may have a higher basal (or minimal) metabolic rate (for given conditions of temperature and alimentation) than the non-fissiparous species. Allen (1919), Hyman (1919) and Petrucci (1962) have shown that minimal metabolic rates exist for several planarian species.

Hyman (1920) has shown that in some planarian species the low 'basal' or 'standard' metabolic rate of early starvation (for *D. dorotocephala*, days 3–14 after last feeding) gives way to an increased metabolic rate associated with 'self-digestion'. After prolonged starvation (about 7 weeks for *D. dorotocephala*) this increased rate becomes comparable with that obtained by feeding. As starvation proceeds at different rates (percentage of weight loss per day or percentage of mg N loss per day) in different species, the duration of the 'standard' metabolic rate cannot be expected to be constant for all species. In fact, the data for *Polycelis nigra* (Hoff-Jørgensen, Lovtrup and Lovtrup 1953) *D. dorotocephala* (Hyman 1919) *Planaria agilis* (Allen 1919, Hyman 1919), and *P. vitta* (Pedersen 1958) yield times for 50 per cent weight (or mg N) loss of 21, 30, 45 and 85 days

127

respectively (my calculations). However, even those species with a short duration of starvation-induced low metabolic rate (and hence low attrition rate) show this minimum during the early post-irradiation period. This period can be expected to be most critical in the determination of survival probability, as shown by Byron and Lajtha (1966) for the rhesus monkey (*Macaca mulatta*).

Hyman (1919) has suggested that the degree of motility of a planarian is related to its metabolic rate. Although Allen (1919) offers data purporting to refute this, statistical analysis of his data reveals a significant metabolic rate decrease ($P < 0.05$) of about 10 per cent. The relative motilities of *D. etrusca* and *D. lugubris*, the former seldom stationary, the latter seldom active, also suggest a higher metabolic rate for the fissiparous species. Thus the starvation-induced decrease in metabolic rate which increased the survival time by 56 per cent in *D. etrusca* also increased the proportion surviving. In contradistinction, lowering the rate for *D. lugubris* resulted in a smaller (22 per cent) increase in survival time, and did not have a beneficial effect on the proportion surviving.

The use of antibiotics which are effective against all organisms commensal to the planarian had no significant effect on the parameters of the planarian x-ray dose–response curve. Thus planarian death after irradiation does not result from infection. The increased survival times following antibiotic treatments suggest that, although regenerative ability was not noticeably affected, a slight decrease in metabolic rate may have occurred. However, as no increase in survival time was observed after combined post-irradiation antibiotics and feeding, any decrease in metabolic rate caused by the antibiotic treatments must have been small compared with the large increase due to feeding.

Planarian death after irradiation appears to be the end-result of excessive depletion of the stock of reproductively intact stem cells (neoblasts), a situation which leads to a lack of tissue renewal and hence eventual organ-failure and death.

ACKNOWLEDGMENTS

I am greatly indebted to Professor Mario Benazzi of the Department of Zoology, and Comparative Anatomy of the University of Pisa, and Dott. F. Garbari, Curator of the Botanical Garden for their kind permission and help in collecting material. Professor Benazzi and Madame Dott. G. Benazzi-Lentati kindly provided 67 tetraploid specimens.

REFERENCES

ALLEN, G. D., 1919, *Am. J. Physiol.*, 49, 403.
ANSEVIN, K. D., and BUCHSBAUM, R., 1961, *J. exp. Zool.*, 140, 155.
BENAZZI, M., 1957, *Caryologia*, 10, 276; 1963, *Monitore zool. ital.*, 71, 288.
BOND, V. P., FLIEDNER, T. M., and ARCHAMBEAU, J. O., 1965, *Mammalian Radiation Lethality* (New York, London: Academic Press).
BRØNDSTED, H. V., 1953, *J. Embryol. exp. Morph.*, 1, 43.
BYRON, J. W., and LAJTHA, L. G., 1966, *Br. J. Radiol.*, 39, 382.
FINNEY, D. J., 1962, *Probit Analysis*, second edition (Cambridge University Press).
GILBERT, C. W., 1968 (in preparation.)
HANSEN-MELANDER, E., MELANDER, Y., and REYNOLDSON, T. B., 1954, *Nature, Lond.*, 173, 354.
HENDERSON, T. R., and EAKIN, R. E., 1959, *J. exp. Zool.*, 146, 253.
HOFF-JORGENSEN, E., LOVTRUP, E., and LOVTRUP, S., 1953, *J. Embryol. exp. Morph.*, 1, 161.

Hyman, L. H., 1919, *Am. J. Physiol.*, **49**, 377; 1920, *Ibid.*, **53**, 299; 1951, *The Invertebrates*: *Platyhelminthes and Rhynchocoela*, Vol. 2 (New York : McGraw-Hill).
Lender, Th., and Gabriel, A., 1961, *Bull. Soc. zool. Fr.*, **86**, 67.
Melander, Y., 1963, *Hereditas*, **49**, 119.
Paterson, E., 1942, *Br. J. Radiol.*, **15**, 302.
Pedersen, K. J., 1958, *J. Embryol. exp. Morph.*, **6**, 308.
Petrucci, D., 1962, *Ricerca scient.*, II–B, **32**, 79.
Rosenbaum, R. M., and Rolon, C. I., 1960, *Biol. Bull. mar. biol. Lab., Woods Hole*, **118**, 315.

STUDIES ON THE CELLULAR BASIS OF RADIATION LETHALITY

In the tables of papers I and II and succeeding papers of this series, the planarian survival curve parameters are listed under the headings LD_{50} and K.

The survival and mortality curves are defined by the same two parameters (number) but

$$K_{survival} = - K_{mortality}$$

with $K_{survival}$ taking negative values only and $K_{mortality}$ taking positive values only.

129

Studies on the cellular basis of radiation lethality
II. Survival-curve parameters for standardized planarian populations

C. S. LANGE

The planarian lends itself to the study of the cellular basis of radiation lethality because it contains a single, morphologically identifiable, population of toti-potential stem cells, without which the animal cannot long survive.

The effects of pre- and post-irradiation decapitation, age (sexual maturity), and hypoxic irradiation conditions are examined to determine further those conditions for which radiation lethality is most likely to be determined solely by the degree of depopulation of reproductively intact stem cells. Under such conditions, whole-animal survival parameters have been determined for each of the five species and polyploid biotypes studied. Radiation lethality was not affected by decapitation or by age, but oxygen enhancement ratios of 2·5 to 3·4 were obtained. At the level of the whole animal, polyploidy gives no protection against radiation. Species differences are at least as great as generic differences in radiosensitivity.

1. Introduction

In an earlier paper (Lange 1968 b) some of the special features of its anatomy and physiology which make the planarian a useful material for the study of the cellular basis of radiation lethality were considered. It was found that the post-irradiation mortality time distribution is uni-modal (a possible indication of a single mortality syndrome) and that the dose-survival curve is sigmoid. Infection was shown not to be a significant factor in radiation mortality, but increased metabolic rate after irradiation was found to be deleterious to survival. Other factors which might affect survival, such as regeneration, age and sexual maturity, as well as a possible dose-dependence of mortality time, are examined in this paper. The question whether planarian radiosensitivity is subject to an ' oxygen effect ', and if so, the magnitude of this effect is also examined. These factors are considered with the aim of determining those conditons under which the degree of depletion of reproductively intact stem cells would be most likely to be the sole determinant of planarian death after irradiation. Under such conditions it might be feasible to attempt quantitatively to relate animal survival to stem-cell survival.

2. Materials and methods

2.1. Materials

The species used in these experiments have been described in an earlier paper (Lange 1968 b).

2.2. Methods

2.2.1. Decapitation

To induce regenerative activity, animals were placed on a block, consisting of a sheet of silicone rubber bonded to a polymethyl methacrylate (Perspex) base, for decapitation. Excess water was drained off, leaving only enough to allow each animal to swim dorsum up in the extended position by ciliary action (not by muscular contractions). Under these conditions, each animal could be observed at $\times 10$ magnification under a Zeiss binocular dissecting microscope and decapitated immediately behind the auricles with a single slicing stroke of a scalpel. A 10A blade (Swan–Morton) gave the best results.

2.2.2. Determination of age and sexual state

For the oviparous species, the age of a planarian reared under defined conditions of temperature and nutrition can be determined from its body length (Reynoldson, Young and Taylor 1965, Lange 1968 a). Methods for the measurement of planarian body length have been described previously (Lange 1967). Sexual maturity comes with the de novo synthesis of the sexual apparatus (Vandel 1920, Fedecka-Bruner 1961) so that sexually mature animals can be selected by examination for the presence of a genital pore. This was done under a Zeiss binocular dissecting microscope at $\times 10$ magnification.

2.2.3. Irradiation

Except for those experiments designed to test for the presence of an 'oxygen effect', all irradiations were carried out as described previously (Lange 1968 b).

For the 'oxygen effect' experiments, all animals to be irradiated (or sham irradiated) were placed in the all-nylon jig shown in figure 1. It consists of a 5 mm-thick nylon 61 annulus of 1·5 in. internal diameter, cut from a 2-in. diameter rod trimmed in such a way that, apart from the points of attachment of the three legs, handle, and closing screw, the walls of the annulus are 3/32 of an inch thick. The legs and handle, with adjustable nylon levelling screws, allow the jig to be accurately positioned ($\pm 0·5$ mm) in the gas-tight pyrex glass irradiation vessel. The top and bottom of the annulus were covered with nylon bolting cloth of $90\,\mu$ pore size and 35 per cent pore cross-sectional area. (14N 'St. Martins' Nylon, Henry Simon Ltd., Stockport, England). Equally good results were obtained with 21N 'St. Martins' Nylon having a $68\,\mu$ pore size and 21 per cent of its cross-sectional area open. The bolting cloth was fused to the annulus with a 20 per cent (w/v) solution of nylon in hot phenol.

After pipetting animals (up to 30 per jig) into the jig, it was sealed with a flush-fitting 2BA nylon screw. The jig was then lowered into the direct-mains water-filled irradiation vessel until all three legs came to rest on the sintered glass filter acting as the bottom of the liquid phase portion of the vessel. Once the jig was in place, the top of the vessel was sealed with a ground-glass plate and vacuum grease. As an additional precaution, bakelite clamps were used to hold the glass plate in place. A photograph of the jig in the irradiation vessel is shown in figure 2. Rapid exchange of water between the volume to which the planarians were confined and the rest of the vessel was confirmed by several dye circulation tests. Air or white-spot nitrogen (British Oxygen Co., < 10 pp $10^6 O_2$) was introduced at the bottom of the vessel through the inlet valve (nylon 7 tubing—

Portex, Portland Plastics Ltd., Hythe, Kent, England—was used to connect the cylinder to the irradiation vessel), passed up through the filter in dense fine streams of bubbles, around the jig, into the air space at the top, out through the outlet valve, through a short length of PVC tubing (Portex, Portland Plastics Ltd., Hythe, Kent, England), into copper tubing, past a bleeder valve, through a flow meter, and through a curved steel tube which could be inserted into a Hersch cell (Hersch 1957 a, b).

Figure 1. Nylon jig used for holding planarians during gassing experiments.

Gassing with white-spot nitrogen at a rate of 40 ml./min reduced the oxygen concentration of the outflow gas to nearly that of the flushing gas after 20 min. (For experiment 4014 this was $6 \cdot 5$ pp $10^6 O_2$; for 5614, ≈ 2 pp $10^6 O_2$; and for 7914, ≈ 10 pp $10^6 O_2$). The background oxygen concentration in the Hersch cell was usually ≈ 1 pp 10^6. A graph of oxygen-clearance rate with gassing time is presented in figure 3. In the experiments reported here, gas-flow rates in excess of 40 ml./min were used and gassing was continued, in all cases, until an oxygen concentration of ≈ 10 pp 10^6 or less was obtained.

When gassing was completed, the cylinder valve was shut, the water level adjusted (lowered) to a level 9·0 mm above the upper layer of bolting cloth, and all three valves were shut. The inflow and outflow connections were removed, and the sealed vessel was placed and centred underneath the x-ray tube head.

Figure 2. The nylon jig in the irradiation vessel.

The apparatus was irradiated from above with a Resomax machine (300 kVp, 2 mm Cu h.v.l., 170 soft tissue rads/min) adjusted so that the distance from the target to the meniscus of the water was 360 ± 1 mm. The x-ray beam was collimated through an 11 cm diameter aperture diaphragm of 6 mm Pb + 2 mm brass. After irradiation, the animals were removed from the jig and handled as in all other irradiation experiments.

3. Results

The process of regeneration is known to stimulate the neoblasts into mitotic activity (Flexner 1898, Lender and Gabriel 1961) and, when completed, to increase metabolic rate, as measured by respiratory activity (Allen 1919, Hyman 1919).

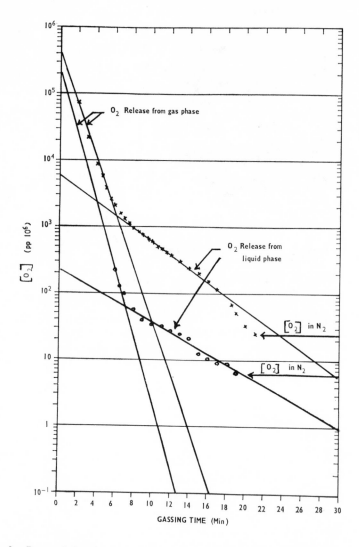

Figure 3. Rates of O_2 clearance. The ordinate denotes the concentration of oxygen (in parts per million) in the white-spot nitrogen which has passed through the irradiation vessel of figure 2. The abscissa shows the duration of gassing with white-spot nitrogen. The two sets of lines are for two different cylinders at two different gas-flow rates. The gas from the 25 pp 10^6 cylinder (upper set of lines) was *not* used for the planarian experiments.

As an increased metabolic rate after irradiation has been shown to be deleterious to planarian survival, the effects of decapitation after irradiation on survival may throw light on the effect of an early increased stem-cell turnover.

3.1. *The effect of decapitation after irradiation*

The survival-curve parameters and the survival times of animals decapitated after irradiation and their controls presented in table 1. Except where otherwise specified, all animals were decapitated 24 ± 3 hours after irradiation and were fasted for the duration of the experiment. No significant changes were found in survival-curve parameters, nor in survival times, for any of the species examined. The one marginal exception to this was the fed *Polycelis tenuis* as compared with the pooled fed control. The decapitated animals were slightly protected ($P \approx 0.05$). In all cases the experimental animals were, prior to decapitation, of the same size as the controls.

Thus we can conclude that decapitation 24 hours after irradiation (24 hours and 48 hours for *Polycelis tenuis*) has little or no significant effect on planarian survival.

3.2. *The effect of decapitation before irradiation*

Animals irradiated at various times after decapitation might be expected to differ in their survival responses, because neoblast mitotic activity and *numbers* are increased in response to the wound (Flexner 1898, Dubois 1949, Lender and Gabriel 1961). The results of decapitation 6 hours and 12 hours before irradiation on *D. etrusca* and decapitation 24 hours and 48 hours before on *D. lugubris* B are shown in table 2. Where the relevant controls had larger than usual uncertainties in the estimation of their survival parameters, the survival parameters of the population standards (combined controls from all control experiments relevant to that species and biotype—see below) are given for comparison.

Animals decapitated before irradiation, in all cases, had slightly increased LD_{50} estimates, but none of the survival-curve parameter changes proved to be significant. The times of death for *D. lugubris* B also did not change significantly. But significant decreases were found in the survival times of *D. etrusca* decapitated 6 hours before ($P < 0.02$ relative to specific control, $P < 0.001$ relative to pooled control) and 12 hours before ($P < 0.02$ relative to pooled control) irradiation.

3.3. *The effect of age and sexual state*

The physiological age of a planarian is closely related to its body size (under given conditions of nutrition and temperature—Abeloos 1930, Lange 1968 a), which is also a determinant of sexual maturity (Fedecka-Bruner 1961, Reynoldson *et al.* 1965). This suggests that metabolic differences may exist between sexually mature (adult) and sexually immature (young) planarians, which may have an effect on survival after irradiation.

The survival-curve parameter and the survival times of young (5–7 mm in length, gential pore absent) as compared with adult (8–12 mm in length, genital pore present) specimens yield no significant differences ($P > 0.05$). The data, however, do suggest slightly lower estimates of LD_{50} and K (mean lethal dose and inverse slope of the probit survival curve) for the young animals (see table 3). Although no significant differences in survival were found, all experiments were carefully controlled for size of animal; as a routine, only adults were used.

	$LD_{50} \pm$ S.E. (rads)	f_s †	$K \pm$ S.E. (rads)	$t_m \pm$ S.E. (days)	f_t ‡	$C \pm$ S.E. (per cent)
D. etrusca						
Control 1401	1090 ± 44	4	219 ± 50	$39 \cdot 8 \pm 1 \cdot 4$	19	$0 \pm 2 \cdot 2$
Decapitated 1406	1178 ± 56	4	228 ± 65	$39 \cdot 8 \pm 0 \cdot 5$	7	$0 \pm 3 \cdot 2$
D. lugubris B						
Control 4703	1772 ± 491	4	1184 ± 1360	$30 \cdot 1 \pm 2 \cdot 9$	13	$0 \pm 8 \cdot 1$
Decapitated 4704	1764 ± 192	4	632 ± 289	$23 \cdot 9 \pm 2 \cdot 0$	14	$0 \pm 5 \cdot 1$
Control 6301	1737 ± 211	6	408 ± 222	$30 \cdot 1 \pm 2 \cdot 1$	9	$22 \cdot 3 \pm 13 \cdot 2$
Decapitated 6303	1658 ± 275	6	698 ± 251	$25 \cdot 6 \pm 1 \cdot 5$	13	$28 \cdot 5 \pm 17 \cdot 6$
Control 7601	1151 ± 497	4	2285 ± 905	$29 \cdot 6 \pm 0 \cdot 8$	23	$15 \cdot 3 \pm 8 \cdot 5$
Decapitated 7603	2447 ± 281	4	1076 ± 436	$31 \cdot 0 \pm 2 \cdot 2$	16	$14 \cdot 2 \pm 7 \cdot 3$
Pooled control § (population standard)	1677 ± 143	49	936 ± 206	$29 \cdot 9 \pm 1 \cdot 0$	47	$12 \cdot 0 \pm 5 \cdot 5$
Pooled decapitated	1940 ± 187	20	910 ± 256	$27 \cdot 0 \pm 1 \cdot 2$	45	$11 \cdot 8 \pm 7 \cdot 1$
D. lugubris A						
Control 7501	2907 ± 93	4	356 ± 126	$29 \cdot 2 \pm 1 \cdot 9$	15	$7 \cdot 5 \pm 2 \cdot 8$
Decapitated 7503	5250 ± 3320	4	2443 ± 2410	$32 \cdot 9 \pm 3 \cdot 5$	6	$0 \pm 3 \cdot 5$
Control 7701	2400 ± 204	6	760 ± 233	$37 \cdot 4 \pm 1 \cdot 9$	23	$0 \pm 3 \cdot 1$
Decapitated 7703	3046 ± 369	4	919 ± 369	$44 \cdot 3 \pm 5 \cdot 6$	5	$0 \pm 2 \cdot 2$
Pooled control § (population standard)	2674 ± 177	18	873 ± 191	$34 \cdot 2 \pm 1 \cdot 5$	39	$4 \cdot 1 \pm 4 \cdot 2$
Pooled decapitated	3620 ± 562	11	1348 ± 472	$38 \cdot 2 \pm 3 \cdot 5$	12	$0 \pm 1 \cdot 9$
Polycelis tenuis						
Control 6201	981 ± 235	5	564 ± 257	$47 \cdot 2 \pm 11 \cdot 6$	5	$8 \cdot 6 \pm 12$
Decapitated 6203	1249 ± 130	3	597 ± 153	$44 \cdot 8 \pm 6 \cdot 7$	8	$0 \pm 3 \cdot 2$
Pooled control (population standard)	706 ± 104	36	388 ± 94	$58 \cdot 4 \pm 4 \cdot 9$	37	$1 \cdot 9 \pm 3 \cdot 1$
Control 7000 ‖	786 ± 52	3	318 ± 63	$38 \cdot 0 \pm 3 \cdot 2$	17	$0 \pm 2 \cdot 2$
Decapitated 7200 ‖	950 ± 82	3	363 ± 101	$46 \cdot 1 \pm 3 \cdot 0$	17	$0 \pm 2 \cdot 3$
Pooled control ‖ (population standard)	718 ± 40	9	272 ± 49	$30 \cdot 6 \pm 2 \cdot 0$	39	$2 \cdot 5 \pm 2 \cdot 7$
48 hours decapitated 7300 ‖	931 ± 62	3	295 ± 77	$32 \cdot 4 \pm 2 \cdot 0$	17	$0 \pm 1 \cdot 8$

Table 1. Survival parameters and survival times for animals decapitated 24 hours after irradiation and their controls.

† Degrees of freedom for survival-curve parameters.
‡ Degrees of freedom for t_m.
§ The survival-curve parameters are estimated from all control experiments, but the t_m is estimated from only those control populations shown in this table.
‖ Experiments 7000, 7200, 7300 and the pooled control were post-irradiation fed experiments.

	$LD_{50} \pm$ S.E. (rads)	f_s	$K \pm$ S.E. (rads)	$t_m \pm$ S.E. (days)	f_t	$C \pm$ S.E. (per cent)
D. etrusca						
− 12 hours	1416 ± 97	2	258 ± 90	$25 \cdot 8 \pm 3 \cdot 2$	16	$0 \pm 3 \cdot 5$
− 6 hours	1526 ± 176	2	547 ± 195	$25 \cdot 2 \pm 2 \cdot 1$	33	$0 \pm 4 \cdot 1$
Control	641 ± 936	2	1361 ± 1140	$31 \cdot 6 \pm 1 \cdot 5$	4	$0 \pm 3 \cdot 2$
Pooled control	1380 ± 14	36	168 ± 14	$34 \cdot 8 \pm 0 \cdot 4$	121	$0 \cdot 1 \pm 0 \cdot 5$
D. lugubris B						
− 48 hours	2067 ± 141	4	458 ± 187	$29 \cdot 0 \pm 1 \cdot 1$	13	$0 \pm 3 \cdot 8$
− 24 hours	1904 ± 95	4	448 ± 127	$27 \cdot 8 \pm 1 \cdot 6$	14	$0 \pm 3 \cdot 2$
Control	1772 ± 491	4	1184 ± 1360	$30 \cdot 1 \pm 2 \cdot 9$	13	$0 \pm 8 \cdot 1$
Pooled control	1677 ± 143	49	936 ± 206	$29 \cdot 9 \pm 1 \cdot 0$	27	$12 \cdot 0 \pm 5 \cdot 5$

Table 2. Survival parameters and survival times for animals decapitated before irradiation and their controls.

	$LD_{50} \pm$ S.E. (rads)	f_s	$K \pm$ S.E. (rads)	$t_m \pm$ S.E. (days)	f_t	$C \pm$ S.E. (per cent)
Polycelis tenuis						
Fed young (7400)	629 ± 35	3	131 ± 36	$36 \cdot 9 \pm 2 \cdot 8$	27	$9 \cdot 3 \pm 4 \cdot 4$
Fed adult (7000)	786 ± 52	3	318 ± 63	$38 \cdot 0 \pm 3 \cdot 2$	17	$0 \cdot 1 \pm 2 \cdot 2$
Pooled fed adults (4400 + 7000)	718 ± 40	9	272 ± 49	$30 \cdot 6 \pm 2 \cdot 0$	39	$2 \cdot 5 \pm 2 \cdot 7$
D. lugubris B						
Young (5219)	1276 ± 112	2	466 ± 130	$25 \cdot 7 \pm 1 \cdot 6$	12	$0 \pm 3 \cdot 2$
Adults (5212)	1400 ± 366	2	400 ± 343	$23 \cdot 8 \pm 1 \cdot 2$	13	$39 \cdot 8 \pm 21 \cdot 1$
Young (6700)	1701 ± 117	4	387 ± 127	$29 \cdot 7 \pm 2 \cdot 5$	18	$0 \pm 4 \cdot 2$
Adults (6800)	1572 ± 2100	4	1005 ± 2890	$30 \cdot 0 \pm 2 \cdot 7$	19	$23 \cdot 1 \pm 29 \cdot 9$
Pooled young (5219 + 6700)	1599 ± 99	9	456 ± 116	$28 \cdot 1 \pm 1 \cdot 6$	31	$0 \pm 3 \cdot 2$
Pooled adults (population standard)	1677 ± 143	49	936 ± 206	$27 \cdot 4 \pm 1 \cdot 7$ †	33	$12 \cdot 0 \pm 5 \cdot 5$

† The survival-curve parameters are estimated from all control experiments, but the t_m is estimated only from those control populations shown in this table.

Table 3. Survival parameters and survival times for sexually immature (young) versus sexually mature adults.

3.4. Relationship between X-ray dose and time of death

3.4.1. The fissiparous species (D. etrusca)

The results of five standard irradiation experiments, differing only in the length of starvation before irradiation (9–20 days), a factor found not to have any significant influence on the survival curve parameters (Lange 1968 b), were examined. It was found by both the Corner test for association (Olmstead and Tukey 1947) and analysis of variance that there was no significant regression of time of mortality on dose. The mean time of death was $34 \cdot 6 \pm 0 \cdot 2$ days (298 animals), which is close to the median value of $34 \cdot 0$ days.

However, the results of three experiments which differed from the above in that an antibiotic (kanamycin, $10 \mu g/ml$.) was present in the post-irradiation milieu, show a significant association between dose and time of death—by the Corner test, $P \approx 0 \cdot 005$, and by analysis of variance $0 \cdot 01 < P < 0 \cdot 05$. The regression equation, obtained by the method of least squares, is found to be:

$$Y = -6 \cdot 1X + 50 \cdot 6 \quad (n = 98 \text{ animals}),$$

where Y is the time of death in days and X is the x-ray dose in krads.

Thus, if mean time of death were to be taken as a measure of rate of decay of functional vital systems after irradiation, then the mean time of death of $34 \cdot 6$ days for animals not given kanamycin in the post-irradiation milieu is equivalent to that caused by an x-ray dose $2 \cdot 6$ krads in kanamycin-treated animals. Similarly, the time of death for animals fed after irradiation is equivalent to $4 \cdot 2$ krads in kanamycin-treated animals; however, feeding after irradiation also affected the fraction surviving.

3.4.2. The oviparous species (D. lugubris and Polycelis tenuis)

Examination of the results of seven standard irradiation experiments on D. lugubris B (in which 265 animals died), in the light of the Corner Test, shows no significant correlation between dose and time of death. The mean time of death was $26 \cdot 5 \pm 0 \cdot 4$ days after irradiation. Similarly, when kanamycin was added to the post-irradiation milieu ($10 \mu g/ml$.), no significant correlation between dose and time of death was found. However, the mean time of death was increased to $47 \cdot 1 \pm 2 \cdot 7$ days after irradiation, a highly significant increase ($P < 0 \cdot 001$).

No significant correlation between time of death and x-ray dose was found for any of the factors investigated in this paper, nor in the previous paper (Lange 1968 b). The same was true for D. lugubris A and Polycelis tenuis.

3.5. The effect of oxygen on planarian radiosensitivity

The effect of irradiation under anoxic conditions was investigated in two genera of planarians, Polycelis felina and D. lugubris B. These studies were made in an attempt to answer the following two questions:

(1) Whether planarian radiosensitivity is subject to an oxygen effect, and if so, the magnitude of this effect in the Planariidae?

(2) Whether the standard conditions of irradiation (Lange 1968 b) used in all x-ray experiments reported in this series are conditions of hypoxia.

The parameters of the planarian survival curves, with their standard errors of estimation, are given in table 4. When the best line was independently fitted to

138

the points of each experiment, there was no case in which the inverse slope of the survival curve obtained under conditions of anoxia was significantly different from that obtained under fully aerated conditions; but the LD_{50}s for the former were in every case highly significantly greater than those for the latter $(P < 0.01)$.

Species	Gas	$LD_{50} \pm$ S.E. (rads)	$K \pm$ S.E. (rads)	$C \pm$ S.E. (per cent)	f_s	Curve fitting constraints
Polycelis felina	N_2	5104 ± 167	728 ± 189	3.1 ± 2.6	5	Expt. No. 5614 none
	N_2	3455 ± 141	1380 ± 1130	45 ± 32	4	Expt. No. 7914 none
	N_2	4447 ± 518	1128 ± 511	12.6 ± 14.5	12	Expt. No. 5614+7914 none
	Air	1954 ± 99	535 ± 118	0 ± 2.7	6	Expt. No. 5601 none
	Air	1131 ± 119	349 ± 102	0 ± 3.2	4	Expt. No. 7901 none
	Air	1535 ± 148	583 ± 176	0 ± 4.2	13	Expt. No. 5601+7901 none
	Air †	1650 ± 126	508 ± 137	8.4 ± 7.4	36	Populatioe standard†
	N_2	3689 ± 554	1175 ± 339	48.5 ± 19.0	9	Expt. No. 7901+7914 Common zero dose intercept OER = 3.29 ± 0.46
	Air	1121 ± 132	357 ± 100	0 ± 3.7		
	N_2	4974 ± 206	1244 ± 149	1.5 ± 2.7	12	Expt. No. 5601+5614 Common zero dose intercept OER = 2.52 ± 0.47
	Air	1971 ± 87	493 ± 59	0 ± 2.8		
D. lugubris B	N_2	4261 ± 193	696 ± 267	0 ± 2.6	4	Expt. No. 4014
	Air †	1677 ± 143	936 ± 206	12.0 ± 5.5	49	Pop .lati n s:a:dard
	N_2	5773 ± 1470	2969 ± 1260	0 ± 5.1	54	Expt. No. 4014+Pop. std. Common zero dose intercept OER = 3.40 ± 0.25
	Air †	1700 ± 135	874 ± 182	13.0 ± 5.4		

† Ungassed dir∍ct mains water.

Table 4. Animal survival-curve parameters for irradiations in N_2 and in air.

However, when the fitted lines for anoxia and fully aerated conditions are constrained to have a single common zero-dose intercept (corrected for natural mortality), then the ratio of LD_{50}s is equal to the ratio of inverse slopes and is highly significantly different from 1.0. When log doses are used and the inverse slopes are constrained to be equal (parallel), almost identical results are obtained. Thus planarian radiosensitivity is subject to an 'oxygen effect' with oxygen enhancement ratios (OER)† of 2.90 ± 0.44 at LD_{50} for *Polycelis felina*‡ and

† OER at $LD_{50} = \dfrac{LD_{50}(N_2)}{LD_{50}(\text{Air})}$ for independent fitting.

$$\text{OER} = \frac{LD_{50}(N_2)}{LD_{50}(\text{Air})} = \frac{K(N_2)}{K(\text{Air})} \text{ for dose-modifying constraints.}$$

‡ For the combined data of experiments 56 and 79. The LD_{50}s for both aerated and anoxic groups in experiment 56 are significantly different from those obtained in experiment 79; however, the OERs calculated for each experiment are not significantly different from each other or from that derived from the combined data of both experiments.

2.54 ± 0.25 (at LD_{50}) or 3.40 ± 0.25 (constrained) for *D. lugubris* B. Furthermore, the animal survival-curve parameters for fully aerated conditions of irradiation (obtained from the special experiments described in this paper) were not significantly different from those of the 'population standards' (see § 3.6), obtained by the normal techniques, described previously (Lange 1968 b).

3.6. Results for standard populations

On the basis of the results of the experiments described in this and in the previous paper (Lange 1968 b), the following conditions were chosen as showing least heterogeneity of response and where planarian death after irradiation would be most likely to be solely determined by depletion of reproductively intact neoblasts. Adult animals, 8–12 mm long, were kept in direct mains water at $20°c$ and subjected to continuous starvation (until well past the endpoint time) starting at least 1 week before irradiation. For the fissiparous species (*D. etrusca, D. tigrina* and *Polycelis felina*) control of animal size was not practicable but animals of a mean length of 7 mm were used.

For each of the five species investigated, the results of experiments performed under these conditions were combined to yield the parameters of the standard population response to 300 kVp x-rays. These 'population standards' are presented in table 5.

4. Discussion and conclusions

The object of this paper has been to investigate those conditions likely to affect radiation-induced mortality probability. It has been previously proposed (Lange 1968 b) that post-irradiation mortality probability depends on two factors: the degree of depletion of reproductively intact stem cells, and the rate of depopulation of essential differentiated tissues (post-irradiation metabolic rate). This latter factor is reflected in the measure of survival time after irradiation. The effects of feeding after irradiation, which alters the metabolic rate, were found to be consistent with this hypothesis.

The effect of decapitation after irradiation was examined to determine if early increased stem-cell turnover has any affect on survival. Løvtrup (1953) found that *D. lugubris* showed no change in metabolic rate during the first 6 days after decapitation, and Lender and Gabriel (1961) have shown that this period is one of intense neoblast mitotic activity. This suggests that either the process of head regeneration has a negligible effect on the metabolic rate of the planarian, or that the decrease in metabolic activity due to decreased movement of the decapitated animal (see Lange 1968 b) is sufficient to offset the increase due to regenerative activity. Both of these possibilities are consistent with the observation that decapitation after irradiation has no effect on the survival time. The finding that decapitation 24 hours after irradiation has little or no significant effect on mortality probability indicates that the rate of stem-cell turnover during this early period after irradiation has little effect on the rate of replacement of radiation-sterilized stem cells. This is not surprising in view of the fact that such neoblasts are capable of differentiation for a limited time (Chandebois 1965, Lange, in preparation) and are thus available for tissue repair and/or head regeneration.

Experiments involving decapitation before irradiation were performed to determine the effect of irradiating actively multiplying neoblasts, on survival. The

140

Species	Biotype	LD$_{50}$ ± S.E. (rads)	K ± S.E. (rads)	f_s	Heterogeneity †	C ± S.E. (per cent)	t_m ± S.E./f_t (days)
D. lugubris	A	2674 ± 177	873 ± 191	18	+ + +	4·1 ± 4·2	33·3 ± 1·0/73
	B	1677 ± 143	936 ± 206	49	+ + +	12·0 ± 5·5	27·2 ± 0·5/246
	D	2141 ± 308	1180 ± 335	5	—	8·8 ± 9·1	35·3 ± 2·4/109
D. etrusca	1	1207 ± 30	269 ± 27	28	—	1·1 ± 1·2	38·0 ± 0·4/231
	2	1380 ± 14	168 ± 14	36	—	0·1 ± 0·6	34·6 ± 0·2/297
D. tigrina		977 ± 55	148 ± 54	20	+ + +	3·9 ± 2·8	38·0 ± 0·4/60
Polycelis tenuis	Fed	718 ± 40	272 ± 49	9	—	2·4 ± 2·7	30·2 ± 1·4/80
	Starved ‡	706 ± 104	388 ± 94	36	+ +	1·9 ± 3·1	38·6 ± 1·7/502
Polycelis felina		1650 ± 126	508 ± 137	36	+ + +	8·4 ± 7·4	37·5 ± 0·9/226

† Non-systematic heterogeneity. —, $P > 0.05$; +, $0.05 > P > 0.01$; + +, $0.01 > P > 0.001$; + + +, $P < 0.001$. All standard errors are corrected for this heterogeneity.

‡ Endpoint time = day 100 after irradiation.

Table 5. Population standards–Survival-curve parameters.

absence of an effect from decapitation 24 hours and 48 hours before irradiation in *D. lugubris* B suggests that neoblasts irradiated while in cell-cycle are not very different in their radiosensitivity from neoblasts at rest. This observation does not, however, preclude the possibility that a difference exists, since only some 10 per cent of the non-blastemal neoblasts are in cycle after decapitation. This proportion is calculated from the known fraction of the population doubling time of the embryonic neoblasts (blastomere—Melander 1963, Le Moigne 1966) occupied by mitosis, and the proportion of neoblasts in mitosis during regeneration (Lender and Gabriel 1961).

The radiosensitivity of immature planarians has been examined because young planarians have been shown to have higher metabolic rates (Allen 1919, Hyman 1920) and fewer neoblasts (Lange 1967, 1968 a) than older planarians. These factors would suggest a greater radiosensitivity for the young animals. However, they also have a higher tissue density of neoblasts (Lange 1968 a). The effect of age on mortality probability was negligible. The decrease in planarian metabolic rate with body size (or age) is far less than the increase caused by feeding (Allen 1919, Hyman 1920). As feeding after irradiation (Lange 1968 b) does not alter the metabolic rate of the oviparous species (especially *D. lugubris*) sufficiently to affect mortality probability, the age-dependent change in metabolic rate is even less likely to do so. Therefore the absence of significant radiosensitivity differences between young and old *D. lugubris* would be compatible with the assumption that *neoblast tissue density* (which decreases with age) and the *absolute number of neoblasts* (which increases with age) are factors of equal importance in the determination of radiation-induced mortality probability.

The effect of oxygen on planarian radiosensitivity is comparable to that demonstrated in many other plant and animal species (cf. Gray 1960). The finding that the animal survival-curve parameters for the specially controlled aerated conditions of irradiation were not significantly different from those of the 'population standards' allows us to conclude that the survival parameters of the 'population standards' result from fully aerated irradiation conditions.

Thus, the results of my experiments are consistent with the conclusion that under the conditions of this study the degree of loss of stem-cell reproductive integrity is the sole determinant of radiation-induced mortality probability.

'Population standards' (i.e. survival parameter values) have been obtained for several species and biotypes. Within the *D. lugubris* polyploid series, the LD_{50} of the diploids is highly significantly greater than that of the triploids. The inverse slopes, however, are not significantly different. Thus the diploid planarian is *more resistant* to radiation-induced mortality than the triploid planarian. The tetraploid sensitivity seems to be between that of the diploid and the triploid. But the variance of the tetraploid LD_{50} estimate is such that no significant difference was found between diploid and tetraploid, or tetraploid and triploid sensitivities. Between species, radiosensitivity increases in the order: *D. lugubris*, *Polycelis felina*, *D. etrusca*, *D. tigrina*, *Polycelis tenuis*. This implies that species differences are at least as important as generic differences in the determination of planarian radiosensitivity.

REFERENCES

ABELOOS, M., 1930, *Bull. biol. Fr. Belg.*, **64**, 1.
ALLEN, G. D., 1919, *Am. J. Physiol.*, **49**, 403.
CHANDEBOIS, R., 1965, *C. r. hebd. Séanc. Acad. Sci.*, *Paris*, **260**, 4834.

DUBOIS, F., 1949, *Bull. biol. Fr. Belg.*, **83**, 213.
FEDECKA-BRUNER, B., 1961, *Archs Anat. microsc. Morph. exp.*, **50**, 221.
FLEXNER, S., 1898, *J. Morph. Physiol.*, **14**, 337.
GRAY, L. H., 1960, *The Initial Effects of Ionizing Radiations on Cells* (London, New York: Academic Press).
HERSCH, P., 1957 a, *Instrum. Pract.*, **11**, 817; 1957 b, *Ibid.*, **11**, 937.
HYMAN, L. H., 1919, *Am. J. Physiol.*, **50**, 67; 1920, *Am. J. Physiol.*, **53**, 399.
LANGE, C. S., 1967, *J. Embryol. exp. Morph.*, **18**, 199; 1968 a, *Expl Geronotol.* (in the press); 1968 b, *Int. J. Radiat. Biol.*, **13**, 511.
LE MOIGNE, A., 1966, *J. Embryol. exp. Morph.*, **15**, 39.
LENDER, TH., and GABRIEL, A., 1961, *Bull. Soc. zool. Fr.*, **86**, 67.
LØVTRUP, E., 1953, *J. exp. Zool.*, **124**, 427.
MELANDER, Y., 1963, *Hereditas*, **49**, 119.
OLMSTEAD, P. S., and TUKEY, J. W., 1947, *Ann. math. Statist.*, **18**, 495.
REYNOLDSON, T. B., YOUNG, J. O., and TAYLOR, M. C., 1965, *J. Anim. Ecol.*, **34**, 23.
VANDEL, A., 1920, *C. r. hebd. Séanc. Acad. Sci.*, Paris, **170**, 249.

STUDIES ON THE CELLULAR BASIS OF RADIATION LETHALITY

In the tables of papers I and II and succeeding papers of this series, the planarian survival curve parameters are listed under the headings LD_{50} and K.

The survival and mortality curves are defined by the same two parameters (number) but

$$K_{survival} = - K_{mortality}$$

with $K_{survival}$ taking negative values only and $K_{mortality}$ taking positive values only.

Studies on the cellular basis of radiation lethality III. The measurement of stem-cell repopulation probability

C. S. LANGE and C. W. GILBERT

A model is presented which defines the parameters of the curve for neoblast (stem cell) survival of reproductive integrity in terms of the parameters of the planarian radiation mortality curve. Two constants of the model, namely, the initial number of neoblasts (N) and the repopulation probability (α) for the independent repopulation of the irradiated planarian tissues by surviving neoblasts are defined and, for the species *Dugesia lugubris*, measured. The assumption that the repopulation probability for each neoblast is independent of the number of neoblasts is confirmed experimentally for diploid and triploid neoblasts of exogenous and endogenous origin. The repopulation probability for diploid (endogenous and exogenous) and triploid (endogenous only) neoblasts are not significantly different. Triploid exogenous neoblasts, however, are only 16 per cent as effective as endogenous triploid or exogenous diploid neoblasts. Values of D_0 and $N\alpha E$ (E = extrapolation number) are given for diploid, triploid and tetraploid neoblasts of *D. lugubris*, and for four other species.

1. Introduction

The planarian lends itself to the study of the cellular basis of radiation lethality because it contains a single, morphologically identifiable, population of totipotential stem cells (cf. Brøndsted 1955), without which the animal cannot long survive. The parameters of planarian mortality have already been established (Lange 1968 c) under conditions where the only likely cause of death was a cellular mechanism, namely, loss of neoblast (stem cell) reproductive integrity. The values of these parameters for such standardized conditions were called 'population standards'. The rescue of supra-lethally-irradiated planarians by added stem cells is investigated below. To deduce the parameters of the cellular response from the animal survival data, the following model was constructed.

2. The model

Let us assume that the neoblast radiation response (i.e. survival of reproductive integrity) can be described by a survival curve which at large doses tends to the asymptotic form:

$$q = E \exp(-D/D_0), \tag{1}$$

where q is the surviving fraction of reproductively intact neoblasts, D is the radiation dose in rads, and E is the extrapolation number and D_0 the 37 per cent dose-slope of this assumed survival curve. Two other parameters are defined: N, the initial number of neoblasts in the animal ($\approx 10^5$; Lange 1967, 1968 a) and α, the probability that a single surviving neoblast repopulates the animal.

Assuming that the animal dies if none of the neoblasts that survive the irradiation, acting independently, succeed in repopulating the animal, then the mortality probability for the animal is:

$$P_m = (1 - \alpha q)^N, \qquad (2)$$

which, for small αq, can be approximated by:

$$P_m = \exp(-N\alpha q) \qquad (3)$$

and by substitution of equation (1),

$$P_m = \exp(-N\alpha E \exp(-D/D_0)). \qquad (4)$$

The mortality curve defined by equation (4) is sigmoid in shape and is closely similar to the Gaussian sigmoid over the range 5 to 95 per cent mortality. The parameters of equation (4) can be related, with a good approximation, (negligible error with respects to the errors of the animal mortality curves) to the LD_{50} and the slope $(1/K)$ of the probit line by the equations:

$$D_0 = K/1 \cdot 2 \qquad (5)$$

and

$$N\alpha E = \exp((LD_{50}/D_0) - 0 \cdot 65). \qquad (6)$$

Equation (4) can be re-written in the form:

$$y(P_m) = -\ln(\ln(1/P_m)) = D/D_0 - \ln(N\alpha E), \qquad (7)$$

so that $y(P_m)$ is a linear function of dose D. This is analogous to the probit line, which is also linearly related to dose,

$$\text{probit}(P_m) = (D - LD_{50})/K. \qquad (8)$$

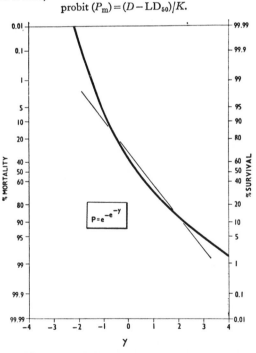

Figure 1. Graph showing the probability plot of the exponential function $P_m = \exp(-\exp(-y))$ (heavy curve) and the probit line (fine straight line) which best fits that curve over the 5–95 per cent region.

145

The relationship between the functions $y(P_m)$ and probit (P_m) is shown in figure 1, where y is plotted against P_m on probability paper. It will be seen that there is only a slight curvature in the range 5 to 95 per cent mortality and so in this range the curve can be replaced by a straight line giving an approximation for y:

$$y(P_m) = 1 \cdot 2 \, (\text{probit} \, (P_m) + 0 \cdot 65). \tag{9}$$

On comparing equations (7) and (8) one obtains the relations:

$$K = 1 \cdot 2 D_0 \tag{10}$$

and

$$\text{LD}_{50} = (\ln \, (N\alpha E) + 0 \cdot 65) D_0. \tag{11}$$

It is of particular interest that the model here described for relating animal mortality probability to cell mortality is very similar to that described by Munro and Gilbert (1961) for relating tumour-mortality probability to cell mortality. This suggests that the mechanisms which relate animal or tumour survival, have much in common.

Thus, from the two parameters of the mortality curve of the whole animal, one can determine D_0 (the inverse slope of the neoblast survival curve) and the product $N\alpha E$. The values of these parameters for each planarian species and biotype are given in table 1.

Species and biotype	Planarian		Neoblast		$\left(\dfrac{\text{S.E.} \, (N\alpha E)}{N\alpha E}\right)$ (per cent)
	$\text{LD}_{50} \pm \text{S.E.}$ (rads)	$K \pm \text{S.E.}$ (rads)	$D_0 \pm \text{S.E.}$ (rads)	$N\alpha E \pm \text{S.E.}$ (rads)	
D. lugubris					
A	2674 ± 177	873 ± 191	727 ± 159	$20 \cdot 66 \pm 15 \cdot 03$	73
B	1677 ± 143	936 ± 206	780 ± 172	$4 \cdot 48 \pm 2 \cdot 39$	53
D	2141 ± 308	1180 ± 335	983 ± 279	$4 \cdot 61 \pm 3 \cdot 74$	81
D. etrusca					
(1)	1207 ± 30	269 ± 27	224 ± 23	114 ± 66	58
(2)	1380 ± 14	168 ± 14	140 ± 12	9970 ± 8060	81
D. tigrina	977 ± 55	148 ± 54	123 ± 45	1470 ± 1715	117
Polycelis tenuis					
(fed)	718 ± 40	272 ± 49	226 ± 40	$9 \cdot 5 \pm 16 \cdot 6$	175
(starved)†	706 ± 104	388 ± 94	324 ± 79	$4 \cdot 62 \pm 3 \cdot 68$	80
Polycelis felina	1650 ± 126	508 ± 137	423 ± 114	$25 \cdot 8 \pm 30 \cdot 6$	119

† Endpoint time = 100 days after irradiation

Table 1. Values of the parameters D_0 and $N\alpha E$ of the neoblast survival curve, and the planarian mortality parameters from which they are derived.

To determine uniquely the extrapolation number of the neoblast survival curve, values of N and α must first be determined.

The number of neoblasts per animal (*D. lugubris*), as found by direct counts, is an allometric function of animal length (Lange 1967, 1968 a). The ' population standards ' (see Introduction), from which the values of table 1 are derived, are based on the results of experiments with planarians 8–12 mm long. For *D. lugubris* of this size range, the number of neoblasts per animal of medium size, N_0, is:

$$N_0 = 1 \cdot 406 \times 10^5,$$

with an upper limit of $1 \cdot 986 \times 10^5$ ($+41$ per cent of median value) and a lower limit of $9 \cdot 18 \times 10^4$ (-35 per cent of median value). The error introduced by deviations from the regression line (Lange 1967, 1968 a) was negligible compared with that resulting from the variation in animal size. If the range is taken to be

146

equivalent to approximately ± 2 standard deviations, the standard deviation (σ) of N_0 is approximately $\pm 0.267 \times 10^5$ or $\approx \pm 20$ per cent.

Thus one can substitute the value of N_0, $(1.406 \pm 0.267) \times 10^5$, for N in the model. Having done so, E can be determined uniquely if the value of α can be measured.

It is possible to measure α directly. According to equation (2) of the model if $q = 1$ as is the case for unirradiated neoblasts, then for unirradiated neoblast grafts, the mortality probability of the supra-lethally radiation-sterilized host is:

$$P_m = (1 - \alpha)^N \qquad (12)$$

or

$$\ln P_m = N \ln (1 - \alpha). \qquad (13)$$

A graph of $\ln P_m$ as a function of N, where N is the number of neoblasts in each graft, should give a straight line of slope $\ln (1 - \alpha)$. For $\alpha < 1$, $\ln (1 - \alpha)$ is a negative quantity, indicating a decreasing host mortality probability with increasing graft size (in terms of numbers of neoblasts).

Alternatively, equation (17) can be written as:

$$\alpha = 1 - \exp (-0.693/N_{50}), \qquad (14)$$

which for large N_{50} becomes:

$$\alpha = 0.693/N_{50}, \qquad (15)$$

where N_{50} is the number of unirradiated neoblasts which must be present in a graft (exogenous or endogenous) to produce a 50 per cent mortality probability in the radiation-sterilized host.

Thus if the neoblasts in a given animal are radiation-sterilized and a defined number of unirradiated neoblasts are grafted (exogenously or endogenously) into such a host, it should be possible to determine α by means of the relationships of equations (13) or (15).

The rest of this paper describes and discusses the results of such grafting experiments.

3. Methods

3.1. *Injection technique*

Jacobson and Jacobson (1963) in an unofficial report to *The Worm Runner's Digest* described a technique for the injection of macromolecules, such as RNA, into the planarian gut. It is known that no digestion takes place within the lumen of the planarian gut (Rosenbaum and Rolon 1960, Jennings 1957) and that cells and macromolecules are incorporated into the planarian via phagocytosis by the gut columnar cells. In the hope that sufficient numbers of neoblasts might be able to penetrate the gastrodermis intact, the following technique was adopted. A single-cell suspension of planarian cells $(\approx 10^6 - 10^7$ cells/ml.) in a 10 per cent solution of Holtfreter's (1931) amphibian and fish saline was injected into the gut of supra-lethally irradiated (10 krads) hosts by means of a siliconized glass micrometer syringe (Agla) with a (Hamilton 35 gauge) 127μ O.D. stainless-steel needle. The needle was inserted into the mouth and up the lumen of the pharynx (which was $\approx 130 \mu$ I.D). After the injection, when the needle was withdrawn, the injected material $(10-30 \mu l. \approx 10^4 - 10^5$ cells) remained in the animal. Material injected directly into the parenchyma is eliminated by the planarian (by contraction) as the needle is withdrawn. What proportion of these cells are neoblasts is not known, but as the neoblasts are small relative to many of the other planarial cells, and are of the order of $< 20 \mu$ diameter (*in vivo*), filtration through a 25μ-pore nylon bolting cloth might be expected to yield a higher proportion than that existing in the animal. However (see §4.1), the injection of such cell suspensions did not increase the survival probability of the supra-lethally irradiated hosts.

3.2. Grafting technique

The grafting technique was a modified version of that of Dubois (1949). For optimal results the grafting area should be cool (10–15°C). These temperatures were achieved by the use of a hood, the inside of which was cooled by blocks of CO_2-ice. The supra-lethally irradiated (10 krads) hosts and the unirradiated donors were placed in 10^{-4}(w/v

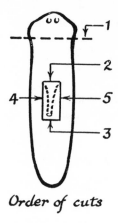

Order of cuts

Figure 2. Order of cuts made in graft host. (1) decapitation; (2) and (3) pre- and post-pharyngeal cuts; (4) and (5) lateral cuts to remove the pharynx.

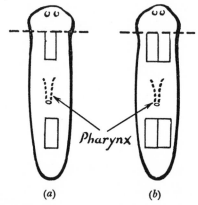

(a)   (b)

Figure 3. Source of grafts. (a) Two hosts grafted from one donor (2H/D), (b) four hosts grafted from one donor (4H/D).

nicotine sulphate solutions for a minimum of 25 to 30 min. This treatment resulted in the contraction and immobilization of the animals. Once immobilized, two hosts and one donor were placed on an ice-cold paraffin block, and all were decapitated with a razor-sharp miniature chisel. The decapitation served the twofold purpose of making the animal less reactive to the subsequent cuttings of the operation, and provided a wound surface some distance away from the implantation site of the graft, so that neoblast migration was encouraged (cf. Dubois 1949). As the major cause of graft failure was the planarian's covering of its wound (cut) surfaces with mucus before the graft and host surfaces could be brought together, the following part of the technique had to be done quickly (2–3 min). For a 1 × 1 mm graft, a chisel with a 1 mm wide blade was used. Using sharp vertically downward strokes, and wiping the blade clean after each stroke, cuts were made in the host as shown in figure 2, and the excised block of tissue removed. As the host contracts somewhat, after the excision, the excised piece must be made a bit larger than 1 × 1 mm. Using two wet paint-brushes (sable hair No. 0 and No. 1) the host was removed from the block and placed dorsum up, on a piece of wet cigarette paper on a nylon drum, the surface of which was common to the miniscus of the 8–10°c water underneath it. The (1 mm chisel was then used to cut out a 1 × 1 mm graft from the donor as shown in figure 3 (*a*) or (*b*), and the graft was implanted into the host with a very fine artists' paint-brush (sable hair No. 00) and a loop of hair. The loop of hair was used to push the host and graft surfaces into contact, and to remove small amounts of mucus if necessary. A lid was placed on the drum-containing tube which was then placed in an 8–10°c waterbath to rest undisturbed in darkness for 3–4 hours. If left any longer, the hosts often dislodged the graft. There-fore, the cigarette paper, with the planarian on it, was placed in a 13 ml. polystyrene tube of direct-mains water (see Lange 1968 b). In this environment, if the planarian moved, it was more likely to propel itself by means of its cilia—a uniform motion less likely to affect the graft adversely. The tube was capped and placed in the 20°c incubator. Two or three days later the paper was removed from the tube and the state of the graft noted. Thereafter, the animal was examined thrice weekly as in most other experiments.

The degree of success depends very much on the planarian population used. Stéphan-Dubois (1966, personal communication) who obtained 98–100 per cent takes with the *D. lugubris* B of Strasbourg, found much less success with the *D. lugubris* B of Nancy. A comparison of our techniques on the latter material yielded very similar results. It is of particular interest that aseptic technique is not necessary for planarian grafting (Stéphan-Dubois 1966, personal communication) and that interspecific grafting is possible (Santos 1931, Lindh 1959).

The number of neoblasts for each size and type of graft was determined directly by means of cell counts on serially sectioned replicate graft pieces (see Lange 1967, for methods).

3.3. *Shielding technique for* $^{90}Sr/^{90}Y$ *β-irradiations*

To test the exstence of a relationship between the number of reproductively-intact neoblasts and the mortality probability of the animal, while leaving the total mass of

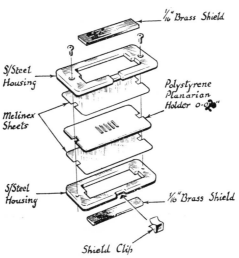

Figure 4. ' Sandwich ' for planarian $^{90}Sr/^{90}Y$ irradiation.

differentiated tissue a constant, parts of the planarian had to be shielded while sterilizing the rest by irradiation. Stéphan-Dubois has done shielding experiments (e.g. Dubois 1949) using 50 kVp x-rays with a lead shield, and Fedecka-Bruner (1964) has sterilized the dorsal neoblasts and testis of *D. lugubris* using 5 kVp x-rays (h.v.l. 0·2–0·3 mm water), while the ventral neoblasts were not sterilized as could be seen from the regeneration which followed.   However, in both these methods the radiation dose to the shielded region must have been at least of the order of 3–10 per cent of the dose given to the sterilized portion, and the dose distribution could not have been uniform (because of attenuation and side scatter).   Although suitable for the experiments for which they were used, these methods would not suffice for an experiment which has as its object the determination of the effects of radiation on the neoblast population.   The following apparatus was therefore designed to irradiate parts of planarians, while shielding with a sharp dose cut-off boundary and eliminating side scatter as a serious contribution to the dose received in the shielded region.

A ⁹⁰Sr source was chosen because of the ease with which ⁹⁰Sr/⁹⁰Y β-particles can be stopped (0·055 in. of stainless steel).   Figure 4 shows the plan of the plastic ' sandwiches ' which hold the planarians and the shields during irradiation, and figure 5 is a drawing of

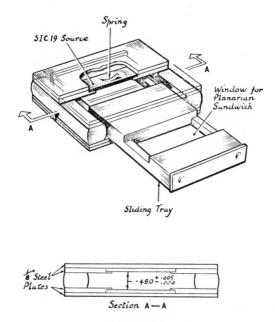

Figure 5.   The ⁹⁰Sr/⁹⁰Y planarian irradiator.

the irradiation device.   No part of those portions which were to be irradiated was situated outside of the 90 per cent isodose ellipse for the window of irradiation.   Figures 6 and 7 show, respectively, the dose drop-off along the short axis of the isodose ellipse with and without the shields in place.   The dose delivered was repeatable to within less ˙ than ± 19·3 rad or 1 per cent, whichever was the greater; while the dose distribution was uniform within less than a 10 per cent total variation.   The dose-rate was 19·3 soft tissue rads/sec.

For a given experiment, the size of the animal (and thus the appropriate cavity to hold that animal), the proportion of the body-length to be shielded (defining the position of the cavity), and the nominal dose to be given to the shielded portion (before applying the shield and supra-lethally irradiating the exposed remainder), were assigned the indices i, j, k respectively. The first two indices defined the appropriate irradiation jig to be used, and the order of irradiation was determined by the sequential appearance of the numbers ijk in a table of random numbers in such a way that the multiplicity of any given ijk had no

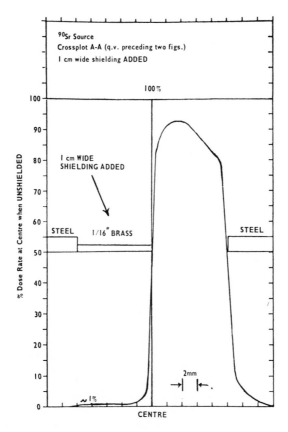

Figure 6. Dose distribution with shield in place. Note that 1 mm from the shield edge, the dose rate is only 5 per cent of peak value; at 2 mm, only 1 per cent. A–A in this figure represents the short axis of the isodose ellipse and is perpendicular to the A–A of figure 5.

effect on its position in the queue. As the whole experiment usually took several days to irradiate, this randomization was of great importance. Irradiations were carried out in a hood where an air temperature of 14–20°c was maintained.

The number of neoblasts in the shielded region was determined by integration of the appropriate neoblast distribution curve(s) (see Lange 1967) over the same region. These curves were determined by direct cell counts of stained serially-sectioned material.

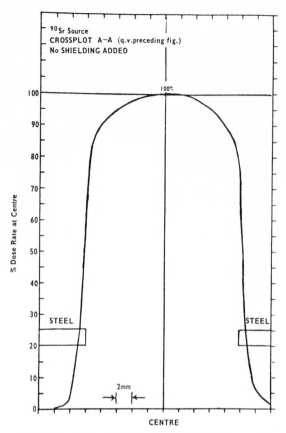

Figure 7. Dose distribution without shield. A–A in this figure is the same as that in figure 6.

4. Results

4.1. *Results of cell suspension injection experiments*

Injection of 46 supra-lethally-irradiated planarians (*D. etrusca*) with cells did not improve survival above that of the saline-injected (10) or uninjected (38) controls. Presumably insufficient numbers of neoblasts get through the gastro-dermis intact (or the cells were damaged by the suspension technique).

To try to determine isotonicity for the planarian cell, specimens of *D. tigrina* were made into cell suspensions in sucrose solutions of 10^{-2} down to 10^{-6} M. Cell diameters were measured, in squashes at $\times 400$ magnification under phase contrast, with an ocular micrometer. As a control value, a planarian was squashed in its own fluids and diameters were measured. The measurements which came closest to the control value were those for 10^{-3} to 10^{-4} M sucrose. At 10^{-6} M there was much debris and cells could be seen to expand and burst. The recent success of Betchaku (1967) in isolating and culturing (short term) neoblasts *in vitro* may make it possible to inject sufficient numbers of neoblasts to bring about the survival of supra-lethally-irradiated planarians.

Biotype	Graft Size (mm)	Type	Ng† ± S.E. (×10⁴)	Number attempted	Number successful	Number of survivors	Percentage of survivors	Percentage of take	Percentage of No-take survivors
A	1 × 1	4H/D	1·85 ± 0·27	100	35	18	51·5	41/190 = 21·6	3/149 = 2
	1 × 2	4H/D	3·01 ± 0·34‡	90	6	5	83·5		
	1 × 1	2H/D	2·06 ± 0·22	—					
B	0·5 × 0·5	4H/D	1·00 ± 0·11‡	6	0	—	—		
	1 × 0·5	4H/D	1·36 ± 0·15‡	48	5	0	0		
	0·5 × 0·5	2H/D	1·75 ± 0·20‡	7	2	2	100		
	1 × 1	4H/D	1·96 ± 0·04	135	11	1	9·1		
	1 × 0·5	2H/D	2·18 ± 0·25‡	8	0	—	—	89/589 = 15·1	3/500 = 0·6
	1 × 1·5	4H/D	2·58 ± 0·29‡	6	0	—	—		
	1 × 1	2H/D	2·90 ± 0·10	121	20	6	30		
	1 × 2	4H/D	3·19 ± 0·54	99	32	3	9·4		
	1 × 1·5	2H/D	3·62 ± 0·41‡	15	1	1	100		
	1 × 2	2H/D	4·34 ± 0·89	79	6	0	0		
	2 × 2	2H/D	7·15 ± 0·82‡	5	3	1	33		
B	Irradiated grafts			60	9				

† Ng = number of neoblasts per graft.
‡ Estimated by interpolation or extrapolation of measured values. Standard errors were estimated to be ∼11·4 per cent as this was the mean percentage of standard error for the measured values.

Table 2. Results of grafting experiments.

4.2. *Results of grafting and shielding experiments* (D. lugubris)

4.2.1. *Diploid–diploid grafts*

An attempt was made to graft 1×1 mm or 1×2 mm blocks of tissue from 80 donors into 339 supra-lethally-irradiated (10 krads) hosts. Of the 100 1×1 mm grafts, 35 were defined as successful (i.e. fused to host on third day after graft placement) and of these, 18 survived to the endpoint time 60 days after irradiation. Of the 90 1×2 mm grafts, only six were successful, yielding five survivors. The combined yield of successful grafts was 21·6 per cent. Of the 149 unsuccessfully grafted hosts, only three survived (it is possible that some neoblasts were transferred during the short time that the graft remained in contact with the host tissues). A summary of the results of the grafting experiments used to determine the value of α are presented in table 2. Equation (13) was fitted to the data by the maximum likelihood method, from which the value of α for exogenous diploid *D. lugubris* neoblasts was found to be $(4·18 \pm 0·85) \times 10^{-5}$.

4.2.2. *Diploid shielding*

Parts of animals were radiation-sterilized while the remaining portions were shielded. After the irradiation, the neoblasts of the shielded region migrated (as described by Dubois 1949; or, more accurately, were transported as described by Betchaku 1967) into the irradiated region (acting as a graft of endogenous cells) and proceeded to repopulate the 'host' tissues. The results of shielding 20 per cent, 50 per cent, and 100 per cent of the animal are presented in table 3.

Ploidy	Method	α	S.E.(α)	Significance of difference from $3n$ grafts
$2n$	Graft Shield	$4·18 \times 10^{-5}$ $3·42 \times 10^{-5}$	$\pm 0·85 \times 10^{-5}$ $\pm 0·76 \times 10^{-5}$	$P < 0·01$ $P < 0·01$
	Combined	$3·76 \times 10^{-5}$	$\pm 0·56 \times 10^{-5}$	$P < 0·01$
$3n$	Graft Shield	$6·33 \times 10^{-6}$ $4·05 \times 10^{-5}$	$\pm 3·35 \times 10^{-6}$ $\pm 0·83 \times 10^{-5}$	— $P < 0·01$

Table 3. Determination of the population probability constant, α for *D. lugubris*.

Equation (13) was fitted to the data, as above, and the value of α for endogenous diploid *D. lugubris* neoblasts was found to be $(3·42 \pm 0·76) \times 10^{-5}$. This value is not significantly different from that for exogenous neoblasts. Therefore the data for endogenous and exogenous neoblasts were combined to yield a value of α for diploid *D. lugubris* neoblasts of:

$$\alpha_{\text{diploid}} = (3·76 \pm 0·56) \times 10^{-5}.$$

A graph of the mortality probability of diploid *D. lugubris* as a function of the number of reproductively intact neoblasts per animal is shown in figure 8.

154

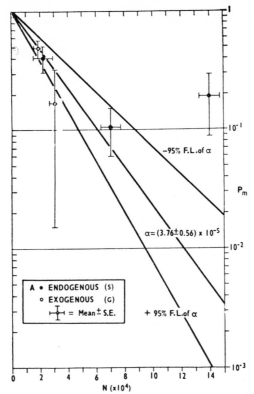

Figure 8. Diploid planarian mortality probability as an exponential function of neoblast number.. $+95$ per cent F.L. and -95 per cent F.L. represent the 95 per cent fiducial limits of α. The point at 14×10^4 neoblasts is not significantly removed from the regression line. (S) = shielded, (G) = grafted, S.E. = standard error.

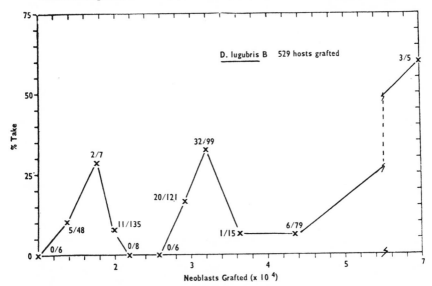

Figure 9. The relationship between percentage-take and number of neoblasts grafted. The fractions represent the number of grafts which fused to the host (took) over the number of animals grafted,

4.2.3. *Triploid–triploid grafts*

Five sizes of graft, cut according to the two or four hosts per donor scheme (see figure 3), made it possible to transplant ten different doses of neoblasts. Of the 584 hosts into which grafts of 0·25, 0·5, 1·0, 1·5, or 2·0 mm² (length × width, thickness approximately constant) were placed, only 86 (14·7 per cent) were successful. No simple relationship was found between graft size and the proportion which were successful (see figure 9). However, equation (13) is consistent with the data (see table 2 and figure 10), yielding a value of α for exogenous triploid *D. lugubris* neoblasts of $(6\cdot33 \pm 3\cdot35) \times 10^{-6}$. This value is highly significantly different from that for exogenous diploid neoblasts ($0\cdot01 > P > 0\cdot001$).

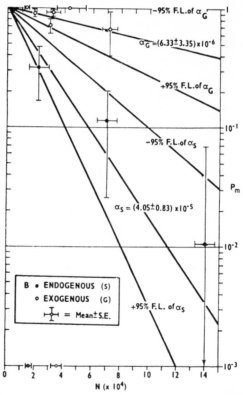

Figure 10. Triploid planarian mortality probability as exponential functions of neoblast number. +95 per cent F.L. and −95 per cent F.L. represent the 95 per cent fiducial limits of α_G and α_S. α_G is derived from experiments using exogenous (grafted) neoblasts while α_S is based on experiments using endogenous (shielded) neoblasts. The two lines are significantly different from each other.

4.2.4. *Triploid shielding*

Experiments identical to those described in §4.2.2 above, were performed on triploid *D. lugubris*. The results of these experiments are shown in table 3. The fitting of equation (13) to the triploid data resulted in a value of α for endogenous neoblasts of $(4 \cdot 05 \pm 0 \cdot 83) \times 10^{-5}$. This value is not significantly different from that for endogenous diploid neoblasts, but is highly significantly different from that for exogenous triploid neoblasts $(0 \cdot 01 > P > 0 \cdot 001)$. Thus, for endogenous neoblasts, the probability of a single surviving neoblast repopulating the animal is not affected by ploidy differences.

5. Discussion and conclusions

The simple model presented in this paper relates the mortality probability of the organism (planarian) to the number, repopulation probability, and two parameters of the assumed multi-target type exponential curve for the survival of reproductive integrity, of the neoblasts (stem cells) of that organism. The success or failure of this model can be seen to lie in the validity of the following four assumptions:

(1) That the mortality probability was measured under conditions in which death was, fundamentally, a result of stem-cell sterilization, and factors which alter the rate of demand for stem-cell differentiation and multiplication were controlled.

(2) That over the range of doses used to determine the organismal mortality probability, the dose-response curve for the survival of neoblast reproductive integrity is not significantly different from the asymptotic exponential portion of a multi-target type curve.

(3) That the survival of reproductive integrity and the repopulation probability for each neoblast is independent of the number of neoblasts at risk.

(4) That factors other than those considered in the model are negligible (under the experimental conditions employed).

The validity of the first and fourth assumptions has been established by the experiments of the first two papers of this series (Lange 1968 b, c) and is further supported by the results of the grafting and shielding experiments of this paper. The validity of the second assumption rests primarily on the observation that even for direct *in vitro* studies, the biological data over the first two or three decades of the cell survival curve are insufficiently precise to distinguish between the two classical two-parameter cell survival curves (multi-hit target and multi-target) (Porter 1967, personal communication). The validity of the third assumption has been demonstrated by the experimental confirmation of the predicted exponential relationship of equation (13). Thus the above model must be deemed successful to the extent that it describes the relationship between the cellular and the organismal radiation response to within the limits of accuracy of the biological data available. The experimental verification of another prediction of the model will be examined later. The recent success of Betchaku (1967) in isolating and maintaining mitotic *in vitro* cultures (albeit short term) of planarian neoblasts may provide the means for a direct measurement of stem-cell reproductive integrity, and possibly a means of studying the effect of the organismal metabolism on the dynamics of *in vivo* post-irradiation recovery and repair.

The difference between the diploid and triploid *D. lugubris* response to grafts of unirradiated tissue may be related to their different growth patterns (Lange 1967), or as is more likely, may be due to the fact that the diploids form a homogenous population whereas the triploids are more heterogenous and may be subject to local population variations similar to those seen between the *D. lugubris* B of Strasbourg and Nancy (Stéphan-Dubois, 1966, personal communication).

Thus the 16 per cent repopulation efficiency ($\alpha_{exogenous}/\alpha_{endogenous}$) of triploid exogenous with respect to endogenous neoblasts may reflect the existence of some six or more physiological phenotypes (nor necessarily sharply defined) such that the neoblasts of a given type could repopulate (as efficiently as endogenous neoblasts) only when grafted into a host of the same phenotype (six types) and/or possibly with a decreased efficiency in the other phenotypes (> 6 types). It may be worth noting that the ability of a graft to take and yield a mosaic, with sizeable distinct and perhaps independent regions, does not necessarily mean that the cells of that graft are able to function in the tissues of the host. Thus the interspecific grafts of Santos (1931) and Lindh (1959) were, as was shown by Lindh (1959), mosaics *but not chimeras*.

It is interesting that the values of the repopulation probability constant, α, derived by both exogenous (grafting) and endogenous (shielding) methods are in such good agreement for the diploids. Although some of the neoblasts in the grafts were likely to have been used to help fuse the graft to the host, the cutting and decapitation of the hosts entailed by the grafting procedure should have given a greater stimulus for neoblast migration from the graft into the depopulated tissues. Thus the magnitudes of these effects are likely to be small, or nearly equal and opposite so that, in the diploid animals, they cancel each other.

Equations (5) and (6) of the model (presented above) reveal two interesting relationships. Firstly, the inverse slopes of the cell and organism survival curves (D_0 and $-K$) are directly related through a normalizing factor (1·2). Thus if all other sources of variation (in cause of death) are controlled, the slope of the stem-cell survival curve is the major (86 per cent) determinant of the slope of the organismal survival curve. Secondly, the intercept of the animal survival curve (LD_{50}) is equal to the product (constant for any given survival curve) $D_0 \ln (E')$, where E' is a modified extrapolation number related to E by the factor $1·9N\alpha$. The 1·9 ($=\exp (0·65)$) is a normalizing factor. In the familiar curve for the survival of reproductive integrity, E is the extrapolated zero-dose intercept under conditions where the zero-dose abcissal value is normalized to unity. Here, $N\alpha$ represents the proportional excess of stem cells above that necessary for the survival level chosen as an endpoint and this is analogous to the cellular multiplicity correction factor introduced by Elkind and Sutton (1960).

The immediate conclusions to be drawn from the values of D_0 and $N\alpha E$ presented in table 1 are firstly, that the neoblasts of *D. lugubris* are much less sensitive to radiation dose increments (i.e. have a higher D_0) than the neoblasts of any other species studied. Secondly, the precision of D_0 is about 10 per cent but the estimate of $N\alpha E$ is only good to about a factor of 2.

In recapitulation, the assumptions inherent in the above model have been validated experimentally and the parameters of the neoblast (stem cell) survival (of reproductive integrity) curve can be determined with reasonable accuracy from the data for the whole animal.

REFERENCES

BETCHAKU, T., 1967, *J. exp. Zool.*, **164**, 407.
BRØNDSTED, H. V., 1955, *Biol. Rev.*, **30**, 65.
DUBOIS, F., 1949, *Bull. biol. Fr. Belg.*, **83**, 213.
ELKIND, M. M., and SUTTON, H., 1960, *Radiat. Res.*, **13**, 356.
FEDECKA-BRUNER, B., 1964, *C. r. hebd. Séanc. Acad. Sci., Paris*, **258**, 3353.
HOLTFRETER, J., 1931, *Arch. EntwMech. Org.*, **124**, 404.

JACOBSON, R., and JACOBSON, A.; as described by McCONNELL, J. V., 1963, *The Worm Runner's Digest*, **5**, 2.

JENNINGS, J. B., 1957, *Biol. Bull. mar. biol. Lab., Woods Hole*, **112**, 63.

LANGE, C. S., 1967, *J. Embryol. exp. Morph.*, **18**, 199; 1968 a, *Expl. Gerontol.* (in the press) 1968 b, *Int. J. Radiat. Biol.*, **13**, 511; 1968 c, *Ibid.*, **14**, 119.

LINDH, N. O., 1959, *Ark. Zool.*, **12**, 183.

MUNRO, T. R., and GILBERT, C. W., 1961, *Br. J. Radiol.*, **34**, 246.

ROSENBAUM, R. W., and ROLON, C. I., 1960, *Biol. Bull. mar. biol. Lab., Woods Hole*, **118**, 315.

SANTOS, F. V., 1931, *Physiol. Zoöl.*, **4**, 111.

Studies on the cellular basis of radiation lethality
IV. Confirmation of the validity of the model and the effects of dose fractionation

C. S. LANGE

The model of Lange and Gilbert (1968) which quantitatively relates the parameters of the planarian survival curve to those of the neoblast (planarian stem cell) survival curve also makes two non-trivial predictions. The first, that planarian mortality probability should decrease exponentially as the number of reproductively-intact neoblasts is increased (total tissue volume being kept constant), has already been verified. The second, that LD_{50} should decrease as a logarithmic function of the number of reproductively-intact neoblasts (total tissue volume being kept constant), is tested and confirmed. The RBE of $^{90}Sr/^{90}Y$ β-rays relative to 300 kVp x-rays was found to be 0·7 to 0·8 for $D. lugubris$. It is also shown that the planarian does not have a superfluity of stem cells, but only enough to ensure survival.

Dose fractionation studies show that the neoblasts of $D. etrusca$ can repair some sub-lethal damage within hours after irradiation and/or are triggered into cell cycle, some phases of which are relatively resistant as compared with the resting state.

1. Introduction

The planarian lends itself to the study of the cellular basis of radiation lethality because it contains a single, morphologically identifiable, population of totipotential stem cells (cf. Brøndsted 1955), without which the animal cannot long survive. The parameters of planarian mortality have been established (Lange 1968 b, c) under conditions where the only likely cause of death was a cellular mechanism, namely, loss of neoblast reproductive integrity. Lange and Gilbert (1968) proposed a model quantitatively relating these parameters (for the organism) to the parameters of the cellular response (neoblast = stem cell, survival of reproductive integrity), measured the constants and coefficients, and tested one prediction of the model.

In this paper, a second prediction of the model (that LD_{50} should vary as a defined function of the number of neoblasts) is examined and tested and the effects of dose fractionation are studied and interpreted in terms of the model.

2. Materials and methods
2.1. Materials

All planarian species and biotypes used in this study have been described previously (Lange 1968 b). Biotype A of $Dugesia\ lugubris$ is diploid, and biotype B is triploid.

2.2. Methods

All the techniques used in this investigation have been described in the previous papers of this series: x-irradiations, see Lange (1968 b); β-irradiations, Lange and Gilbert (1968); decapitations and other trans-sections, Lange (1968 c).

3. Results

3.1. A test of a prediction of the model

One of the relationships predicted by the model of Lange and Gilbert (1968; see equation (11)) is that:

$$LD_{50} = D_0(\ln (N\alpha E) + 0.65),$$

where N is the initial number of neoblasts, α is the independent repopulation probability for a single neoblast and D_0 and E are the inverse slope and extrapolation number of the neoblast survival curves. As values of N, α, D and E have been obtained (for the first three see Lange 1967, 1968 a, Lange and Gilbert, 1968; and for the last, Lange 1968 d), and if D_0, α, and E are independent of N, then it should be possible to predict changes in LD_{50} following given changes in N.

Three methods for varying N were considered, namely, (1) cutting the irradiated ($LD_{50} \pm LD_{50}$ doses) planarian into pieces each of which contained N_i neoblasts, (2) shielding portions of the planarian during the radiation ablation (10 krads) of the remainder, so that the shielded region contained N_i neoblasts (and then giving doses in the $LD_{50} \pm LD_{50}$ range to the shielded region), and (3) grafting irradiated ($LD_{50} \pm LD_{50}$ doses) pieces containing N_i neoblasts into radiation-ablated (10 krads) hosts. Owing to the low efficiency of the grafting method, sufficient data for accurate LD_{50} determinations were obtained only from the first two methods.

The first set of experiments involved cutting planarians 24 hours after irradiation to produce pieces containing 16 per cent (20 per cent by length), 40, 60 and 84 per cent (80 per cent by length) of the planarian's neoblasts (the planarian's neoblast distribution, see Lange 1967, 1968 a, is such that the average number of neoblasts per unit length in the anterior and posterior 20 per cent of the animal is less than that for longer pieces). The LD_{50} for each of these size groups was compared with that for the other size groups and with the predicted value. Pieces of the same size group but differing in cephalo-caudal origin were treated as separate sub-groups. The results of these experiments on *D. lugubris* A and B and *Polycelis tenuis* are presented in tables 1 and 2 and are compared (*D. lugubris* only) with the predicted values in table 3.

The net result was that no systematic or significant change in LD_{50} was observed related to size of piece or to cephalo-caudal origin. These results do not agree with the predicted decrease in LD_{50} with decrease in N, but large variances of the predicted values (the result of accumulated uncertainties) render the disagreement not significant. However, as we shall see below, the absence of a difference in these experiments should have been expected and is a positive datum in support of the model.

The second set of experiments altered N by shielding a variable given portion of each animal during the radiation sterilization of the remainder. This allows a decrease in N without an alteration in total tissue volume.

161

Species	LD$_{50}$ ± S.E. (rads)	K ± S.E. (rads)	C ± S.E. (per cent)	f_s	Size (per cent)
D. lugubris A	2674 ± 177	873 ± 191	4·1 ± 4·2	18	100
	2994 ± 173	428 ± 231	3·4 ± 2·1	11	78 A†
	3620 ± 562	1348 ± 472	0·0 ± 1·9	11	78 P†
	2531 ± 240	894 ± 275	0·0 ± 2·7	11	22 A
	2752 ± 283	1004 ± 368	4·9 ± 4·0	11	22 P
	2674 ± 177	873 ± 191	4·1 ± 4·2	18	100
	3083 ± 141	348 ± 187	4·8 ± 1·8	25	78
	2820 ± 171	924 ± 207	2·3 ± 2·3	25	22
D. lugubris B	1677 ± 143	936 ± 206	12·0 ± 5·5	49	100
	1862 ± 198	714 ± 208	26·7 ± 8·7	11	80 A
	2008 ± 246	941 ± 310	16·0 ± 9·3	11	80 P
	1803 ± 114	327 ± 118	3·2 ± 4·4	6	60 A
	1579 ± 112	263 ± 105	3·7 ± 6·6	6	60 P
	1875 ± 284	898 ± 374	0·0 ± 5·7	6	40 A
	2021 ± 207	648 ± 253	2·0 ± 6·2	6	40 P
	1592 ± 268	809 ± 294	15·4 ± 11·4	11	20 A
	1987 ± 376	1335 ± 567	21·7 ± 10·8	11	20 P
	1677 ± 143	936 ± 206	12·0 ± 5·5	49	100
	1903 ± 181	896 ± 222	19·7 ± 6·7	29	80
	1675 ± 85	294 ± 90	3·1 ± 4·2	15	60
	1935 ± 206	833 ± 288	0·0 ± 5·6	15	40
	1757 ± 226	1043 ± 281	18·7 ± 8·1	29	20

† A = anterior fragment; P = posterior fragment.

Table 1. Effect of size of planarial fragment on survival (*D. lugubris*). (K = inverse slope of probit regression line, C = natural mortality, f_s = degrees of freedom of the survival curve parameters, LD$_{50}$, K and C.)

Species	LD$_{50}$ ± S.E. (rads)	K ± S.E. (rads)	C ± S.E. (per cent)	f_s	Size (per cent)
Polycelis tenuis	981 ± 235	564 ± 257	8·6 ± 12·1	5	100
	999 ± 239	706 ± 291	9·8 ± 11·9	3	80 A†
	1249 ± 130	597 ± 153	0·0 ± 3·2	3	80 P†
	919 ± 110	506 ± 130	0·0 ± 3·2	3	60 A
	1121 ± 141	521 ± 148	0·0 ± 4·1	3	60 P
	781 ± 222	579 ± 245	11·1 ± 13·3	3	40 A
	979 ± 184	930 ± 325	0·0 ± 3·2	3	40 P
	—	—	—	—	20 A
	—	—	—	—	20 P
	706 ± 104	388 ± 94	1·9 ± 3·1	36	100
	1140 ± 121	661 ± 143	5·0 ± 5·3	9	80
	1017 ± 78	513 ± 89	0·0 ± 2·2	9	60
	863 ± 152	794 ± 216	5·0 ± 5·6	9	40
	—	—	≈ 5	—	20

† A = anterior fragment; P = posterior fragment.

Table 2. Effect of size of planarial fragment on survival (*Polycelis tenuis*). (Symbols as in table 1.)

Biotype	N_0	α	E	NαE	Percentage of N intact†	D_0	ln (NαE)	Predicted‡ X-ray LD$_{50}$	Observed X-ray LD$_{50}$	RBE§	Predicted‖ β-LD$_{50}$	Observed β-LD$_{50}$
A	$(1{\cdot}406\pm0{\cdot}267)\times10^5\times1{\cdot}00$	$(3{\cdot}76\pm0{\cdot}56)\times10^{-5}$	$3{\cdot}9\pm3{\cdot}0$	$20{\cdot}66\pm15{\cdot}03$	100	727 ± 159	$3{\cdot}030$	2675 ± 715	2674 ± 177	$0{\cdot}805\pm0{\cdot}090$	3323 ± 961	3320 ± 176
	,, ×0·84	,,	,,	$17{\cdot}36\pm13{\cdot}64$	84	,,	$2{\cdot}856$	2549 ± 729	3083 ± 141	,,	3166 ± 971	—
	,, ×0·60	,,	,,	$12{\cdot}42\pm10{\cdot}12$	60	,,	$2{\cdot}518$	2303 ± 715	—	,,	2861 ± 943	—
	,, ×0·50	,,	,,	$10{\cdot}33\pm8{\cdot}72$	50	,,	$2{\cdot}337$	2172 ± 716	—	,,	2698 ± 938	3673 ± 248
	,, ×0·40	,,	,,	$8{\cdot}28\pm7{\cdot}35$	40	,,	$2{\cdot}114$	2009 ± 729	—	,,	2406 ± 946	—
	,, ×0·16	,,	,,	$3{\cdot}31\pm4{\cdot}64$	16	,,	$1{\cdot}197$	1343 ± 1034	2626 ± 171	,,	1068 ± 1300	465 ± 1701
B	$(1{\cdot}406\pm0{\cdot}267)\times10^5\times1{\cdot}00$	$(4{\cdot}05\pm0{\cdot}83)\times10^{-3}$	$0{\cdot}79\pm0{\cdot}47$	$4{\cdot}48\pm2{\cdot}39$	100	780 ± 172	$1{\cdot}500$	1677 ± 490	1677 ± 143	$0{\cdot}678\pm0{\cdot}080$	2472 ± 775	2472 ± 200
	,, ×0·84	,,	,,	$3{\cdot}76\pm2{\cdot}28$	84	,,	$1{\cdot}320$	1537 ± 526	1903 ± 181	,,	2266 ± 817	—
	,, ×0·60	,,	,,	$2{\cdot}69\pm1{\cdot}73$	60	,,	$0{\cdot}990$	1279 ± 531	1675 ± 85	,,	1885 ± 844	—
	,, ×0·50	,,	,,	$2{\cdot}24\pm1{\cdot}52$	50	,,	$0{\cdot}806$	1136 ± 547	—	,,	1674 ± 829	2185 ± 223
	,, ×0·40	,,	,,	$1{\cdot}79\pm1{\cdot}31$	40	,,	$0{\cdot}582$	961 ± 582	1935 ± 206	,,	1417 ± 874	—
	,, ×0·16	,,	,,	$0{\cdot}72\pm0{\cdot}93$	16	,,	$-0{\cdot}3285$	251 ± 1012	1757 ± 226	,,	370 ± 1497	717 ± 300

† Percentage remaining after cut (x-ray), or percentage shielded (β-ray). ‡ Predicted LD$_{50}$ = D_0 (ln (NαE)+0·65). § RBE = $\dfrac{\text{X-ray dose}}{\text{β-ray dose}}$. ‖ Predicted β-ray LD$_{50}$ = predicted x-ray LD$_{50}$ ÷ RBE.

Table 3. Predicted and observed LD$_{50}$ values for diploid (A) and triploid (B) D. lugubris. (Symbols N, α, and E are defined in §3.1; N_0 is the value of N for animals of the size used in these experiments.)

To obtain a reasonably sharp dose cut-off at the edge of the shield, ^{90}Sr/^{90}Y β-rays were used (see Lange and Gilbert 1968). About 1 per cent of the 10 krads (in soft tissue) ablation dose penetrated the shields in the form of Bremsstrahlung, but all β-rays were stopped. The RBE of ^{90}Sr/^{90}Y β-rays (with respect to 300 kVp x-rays) is reported in § 3.2 of this paper. The results of these experiments with diploid and triploid *D. lugubris*, in which 80, 50 or 0 per cent of each animal was radiation sterilized, are presented in table 4, and are compared with the predicted values in table 3. As in the cutting experiments, the observed LD$_{50}$ values were not significantly different from the predicted values; but the LD$_{50}$ decreased in the expected manner. For biotype B (triploid) the decrease in LD$_{50}$ of animals 20 per cent of which were shielded, compared with animals 50 or 100 per cent of which were shielded, was found to be highly significant ($P < 0.01$).

Species	LD$_{50}$† ± S.E.	K† ± S.E.	C ± S.E. (per cent)	f_s	Percent shielded
D. lugubris A	3320 ± 176	595 ± 244	$19 \cdot 1 \pm 7 \cdot 2$	27	100
	3673 ± 248	1993 ± 415	$10 \cdot 5 \pm 3 \cdot 4$	27	50
	465 ± 1701	4341 ± 3211	$40 \cdot 0 \pm 10 \cdot 4$	27	20
D. lugubris B	2472 ± 200	967 ± 270	$1 \cdot 9 \pm 3 \cdot 1$	27	100
	2185 ± 223	1055 ± 321	$11 \cdot 4 \pm 7 \cdot 3$	27	50
	717 ± 360	1064 ± 310	$31 \cdot 6 \pm 10 \cdot 9$	27	20

† In units of ^{90}Sr/^{90}Y β-rads in soft tissue.

Table 4. Effect of shielding on survival of *D. lugubris* after ^{90}Sr/^{90}Y β-irradiation. (Symbols as in table 1.)

To confirm that the observed difference between the first and second set of experiments could not be attributed to the effect of cutting *per se*, either by stimulation of migration (Dubois 1949) or of mitotic activity (Verhoef 1946, Lender and Gabriel 1961), a third set of experiments was performed identical in all respects to the second save that immediately after irradiation a piece of tissue the size of the head was removed from the end distal to the shielded region. The results of these experiments are shown in table 5. No significant difference was found between the results of shielding-plus-cutting and shielding only. The results of the shielding-plus-cutting experiments are also consistent with the predicted decrease in LD$_{50}$ in that the LD$_{50}$ of animals 20 per cent of which had been shielded was found to be significantly ($P < 0.05$) less than that of 50 and 100 per cent shielded animals.

3.2. RBE of ^{90}Sr/^{90}Y β-rays with respect to 300 kVp X-rays in D. lugubris

Inspection of the inverse slopes (K's) of the computer fitted planarian mortality probit regression lines (with their standard errors for reliability of estimation), allows one to conclude that there is no significant difference between the inverse slope of diploid (A) and triploid (B) or of x and β-ray probit whole animal mortality curves (table 6). Thus if these four groups of animals do

Treatment (per cent)	Endpoint	LD$_{50}$ ± S.E. (rads)	K ± S.E. (rads)	C ± S.E. (per cent)
0 ablated†	65	3120 ± 298	595 ± 413	19·1 ± 10·3
0 ablated‡ + cut	day survival	3676 ± 432	1129 ± 777	10·5 ± 5·9
50 ablated	65	3673 ± 387	1993 ± 648	10·5 ± 4·6
50 ablated + cut	day survival	4072 ± 953	2246 ± 1670	17·7 ± 8·9
80 ablated	65	466 ± 1987	4341 ± 3751	40·1 ± 10·3
80 ablated + cut	day survival	1891 ± 443	2940 ± 1048	22·0 ± 5·9
0 ablated	100	3152 ± 273	722 ± 403	26·3 ± 9·1
0 ablated + cut	day survival	3735 ± 488	1382 ± 345	19·2 ± 6·8
50 ablated	100	4254 ± 1564	1766 ± 2258	39·5 ± 13·4
50 ablated + cut	day survival	3905 ± 800	2468 ± 1494	26·6 ± 7·9
80 ablated	100	1483 ± 1317	3105 ± 2399	60·1 ± 9·6
80 ablated + cut	day survival	1471 ± 667	3220 ± 1399	32·1 ± 6·8 §

† x per cent of planarian is radiation ablated with 10 krads and then the shielded portion is given a dose in the range LD$_{50}$ ± LD$_{50}$.

‡ x per cent of the planarian is radiation ablated with 10 krads, then the shielded portion is given a dose in the range LD$_{50}$ ± LD$_{50}$, and finally a piece the size of the head is cut from the ablated portion most distal to the shielded portion.

§ No significant difference was found between values for the two treatments except for the natural mortality (C) of the 80 per cent ablated, 100 day survival group.

Table 5. The absence of an effect of a cut (in the radiation ablated region) on the parameters of the planarian mortality curve (*D. lugubris* A). (Symbols as in table 1.)

Radiation quality / Ploidy	x-ray				β-ray			
	K (rads)	S.E. (rads)	Var.	f_s	K (rads)	S.E. (rads)	Var.	f_s
2n	873	191	3·6481 × 10^1	18	595	244	5·9312 × 10^1	27
3n	936	206	4·2436 × 10^1	49	967	270	7·2900 × 10^1	27

Table 6. Inverse slope of survival curves for diploid and triploid *D. lugubris* after x-rays and β-rays. (Symbols as in table 1.)

derive from different populations, the accuracy of the inverse slope determinations does not allow this to be apparent. As K is a measure of the variance of the animals at risk, with respect to survival after irradiation, and as there is no evidence that the value of K differs for each of the four groups, then, if the mean (LD$_{50}$) of each of these four groups does not differ, one may conclude that all the samples derive from a homogeneous population (with respect to survival).

According to the model (Lange and Gilbert 1968):

$$D_0 = K/1 \cdot 2 = 0 \cdot 833K$$

and

$$LD_{50} = D_0 \left(\ln \left(N\alpha E \right) + 0 \cdot 65 \right),$$

so that a change in D_0 is reflected in changes in K and LD_{50}. However, under our experimental conditions, K has a standard error of ± 22 per cent for x-rays(28–41 per cent for β-rays), whereas the LD_{50} has a standard error of only 5 per cent (for diploids) or 8 per cent (for triploids). Thus D_0 changes of the order of 10–20 per cent would not result in significant changes in K, but could be detected by the LD_{50} change (for $N\alpha E$ = constant). Therefore, since the LD_{50} is a statistic of much greater accuracy of determination than K, and as values of N, α or E are not expected to change radically with radiation quality, if the LD_{50}'s of the four groups differ, then it is most likely that these differences are due to changes in D_0. Estimates of the relative biological effectiveness (RBE) of $^{90}Sr/^{90}Y$ β-rays to 300 kVp x-rays for diploid and triploid $D.$ $lugubris$, based on the LD_{50} data given in table 7, are presented (with their standard errors) in table 8.

Radiation quality Ploidy	x-ray				β-ray			
	LD_{50} (rads)	S.E.† (rads)	Var.	f_s	LD_{50} (rads)	S.E.† (rads)	Var.	f_s
$2n$	2674	177	$3 \cdot 1329 \times 10^4$	18	3320	176	$3 \cdot 0814 \times 10^4$	27
$3n$	1677	143	$2 \cdot 0449 \times 10^4$	49	2472	200	$4 \cdot 0000 \times 10^4$	27

† S.E. of LD_{50} estimation; S.E. for all rad doses quoted $= \pm 3$ per cent ($=$r.m.s. of 2 per cent variation +2 per cent deviation from N.P.L. standard.)

Table 7. LD_{50} of diploid and triploid $D.$ $lugubris$ after x-rays and β-rays. (Symbols as in table 1.)

Ploidy	RBE†	S.E.	Var.	f
$2n$	0·805	0·068	$4 \cdot 65 \times 10^{-3}$	45
$3n$	0·678	0·080	$6 \cdot 358 \times 10^{-3}$	76

$$\dagger \ RBE = \frac{LD_{50} \ (\text{x-ray})}{LD_{50} \ (\beta\text{-ray})} = RBE \frac{(\beta\text{-ray})}{(\text{x-ray})}.$$

Table 8. RBE of $^{90}Sr/^{90}Y$ β-rays to 300 kVp x-rays in diploid and triploid $D.$ $lugubris$. (f = degrees of freedom.)

The β-ray/x-ray RBE's are significantly different from 1·0 at the 95 and 99 per cent fiducial levels for diploids and triploids respectively, but the RBE values found for diploids and triploids are not significantly different from each other (by Student's t test after confirming its applicability by an 'F' ratio for equality of variances test).

3.3. *Results of X-ray dose-fractionation experiments*

In order to examine the planarian's capacity to repair radiation-induced sub-lethal damage, dose fractionation experiments were performed (with the species *D. etrusca*).

An initial dose (conditioning dose, D_c) of either 800 or 1000 rads was followed at intervals of 4, 6, 8, 12, 24 or 48 hours by one of a series of doses sufficient to determine planarian mortality curves. The graph of the LD_{50} as a percentage of the single acute dose LD_{50} (control), given in figure 1, shows that even a 4-hour fractionation interval was long enough for significant repair to take place ($P < 0.01$). Although there was no evidence of a significant increase in repair by lengthening the interval between doses from 4 to 24 hours, the slightly lower LD_{50} for the 6 hour-interval group is suggestive of a heavily damped Elkind-type (Elkind and Sutton 1959) recovery curve. The LD_{50} of the 48-hour group was significantly increased above that of the 24-hour group ($0.05 > P > 0.01$) and highly significant above control (single dose) LD_{50} ($P < 0.001$). This increase is thought to result from an increased multiplicity of neoblasts and is thus a measure of repopulation rate rather than cellular repair.

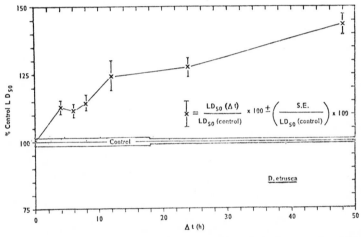

Figure 1. Variation in split dose LD_{50} (as a proportion of single acute dose LD_{50}) with fractionation interval (Δt). Control = single acute dose.

Dose fractionation was also found to change the slope of the survival curve. However, the pattern was not quite that observed for the LD_{50} (see figure 2). With the exception of the 8-hour fractionation interval, the slope of the survival curves for all other dose-fractionation intervals was found to be significantly (at 4, 24 and 48 hours) or highly significantly ($P < 0.01$ at 6 and 12 hours) shallower than that of the control survival curve. The observation that the slope does not return to the control value after 24 or 48 hours suggests that a relatively long-lived heterogeneity of response to the second dose fraction may be induced by the conditioning dose (see table 9). This heterogeneity may be caused by an increase in the proportion of neoblasts in cell-cycle, some parts of which may be more resistant (greater D_0).

167

A graph of the difference between the LD_{50} for each fraction interval and control LD_{50} reveals the pattern of repair of the sub-lethal damage caused by the conditioning dose D_c, or its converse, the pattern of decay of residual injury (i.e. $1 - (LD_{50}/D_c)$). Such graphs of damage repair and injury decay are shown in figures 3 and 4 (see tables 9 and 10 for data). The data suggest a possible exponential decay of residual injury, but further experiments using longer fractionation intervals are necessary to confirm this hypothesis.

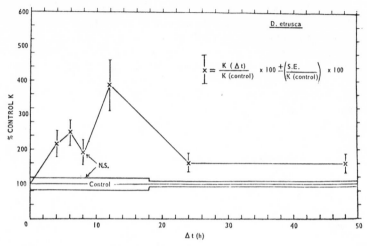

Figure 2. Variation in split dose inverse slope (K) (as a proportion of K for a single acute dose) with fractionation interval (Δt). Control = single acute dose; N.S. = difference not significant.

Δt	$LD_{50} \pm$ S.E. (rads)	$K \pm$ S.E. (rads)	f_s	$C \pm$ S.E. (per cent)	$\bar{t}_m \pm$ S.E. (days)	f_t
Control (4–12 hours)	1314 ± 19	95 ± 17	7	$0 \pm 1\cdot6$	$27\cdot0 \pm 1\cdot0$	48
4 hours	1482 ± 33	206 ± 36	7	$0 \pm 1\cdot6$	$26\cdot7 \pm 1\cdot3$	29
6 hours	1464 ± 34	237 ± 36	7	$0 \pm 1\cdot6$	$25\cdot8 \pm 1\cdot2$	29
8 hours	1504 ± 38	182 ± 35	7	$0 \pm 2\cdot0$	$27\cdot6 \pm 2\cdot0$	20
12 hours	1635 ± 72	368 ± 70	7	$0\cdot2 \pm 2\cdot4$	$24\cdot3 \pm 0\cdot4$	41
24 hours	1759 ± 50	273 ± 45	9	$0 \pm 1\cdot0$	$35\cdot3 \pm 0\cdot6$	25
48 hours	1982 ± 51	269 ± 48	9	$0\cdot5 \pm 1\cdot2$	$36\cdot1 \pm 3\cdot2$	14
Control† (24–48 hours)	1380 ± 14	168 ± 14	36	$0\cdot1 \pm 0\cdot5$	$37\cdot8 \pm 2\cdot0$	21

† The survival curve parameters of the controls were in excellent agreement with those of the relevant population standard, so the standard with its smaller variance was substituted. $\bar{t}_m$, however, is calculated from the controls of these experiments.

Table 9. Results of x-ray dose-fractionation. (f_s = degrees of freedom of LD_{50}, K, and C; $\bar{t}_m$ = mean post-irradiation survival time; f_t = degrees of freedom of t_m.)

For all the fractionation intervals studied, survival time remained constant at the control (single acute dose) value except for the 12-hour group, for which there was a slight but significant shortening (see table 9).

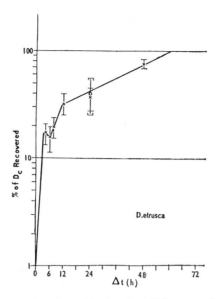

Figure 3. Proportion of conditioning dose (D_c) recovered with time.

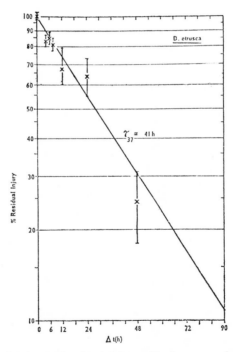

Figure 4. Exponential decay of residual injury with time. τ_{37} = half-life of remaining
residual injury. Percentage residual injury = $[100 - (\Delta LD_{50}/D_c) \times 100]$.

4. Conclusions and discussion

The results of the cutting, shielding and shielding-plus-cutting experiments show that the LD_{50} is independent of N (the initial number of neoblasts) when N is decreased in such a manner that total tissue volume is also decreased; but the LD_{50} is a function of N when N is decreased in such a way that total tissue volume (V) remains constant. Since cutting *per se* does not affect LD_{50} (Lange 1968 c—decapitation; Lange and Gilbert 1968—grafting; and this paper—shielding plus cutting), and as the results of the shielding experiments support the model, the most likely explanation of the discrepancy between the results of the cutting and shielding experiments is that α (the probability that a single surviving neoblast can independently repopulate the planarian) is inversely related to V (i.e. is equal to a function of the reciprocal of V). As planarian differentiated tissue volume (which is not very different from total tissue volume) increases, the demand for stem-cell differentiation to replace worn and/or lost differentiated cells might be expected to increase. Therefore, the value of α can be expected to reflect changes in differentiation demand, in that increased demand must decrease the likelihood of repopulation (from a given number of neoblasts) before vital organ failure and death ensue. Thus the product $N\alpha$ can be seen to be a measure of *neoblast tissue density* (with constant differentiation demand),

Δt	D_c (rads)	$\dfrac{\Delta LD_{50}}{D_c} \times 100$ per cent	R.I. (per cent)	$\pm$ S.E. (per cent)
Control	0	0	100	1·4
4 hours	1000	16·8	83·2	3·7
6 hours	1000	15·0	85·0	3·9
8 hours	1000	19·0	81·0	4·2
12 hours	1000	32·1	67·9	7·4
24 hours	1000	36·0	64·0	·8·8
24 hours	800	39·5	60·5	14·9
48 hours	800	75·3	24·7	6·6

Table 10. Repair of residual injury (R.I.) in *D. etrusca*. D_c, conditioning dose.

a factor which has previously been shown to be of importance in planarian lethality and senescense (Lange 1968 a, c). Although neoblast tissue density was not included in the model as a primary factor, it is accounted for (in the model) by the two factors N and α. Taking the above relationships into account, the results of both cutting and shielding experiments become compatible with the predictions of the model.

The experimental confirmation of two non-trivial predictions of the model, i.e. the radiation LD_{50} variation with N (for constant α) and the exponential decrease of mortality probability with increasing N (Lange and Gilbert 1968), is strong evidence for the validity of the model.

While the product $N\alpha$ can be considered as a measure of neoblast tissue density, it can also be taken as a measure of the planarian's proportional excess of neoblasts above that required to yield a 63 per cent survival probability ($N\alpha = 1$). Thus the $N\alpha$ values of 5–6 found for *D. lugubris* suggest that this species does not have a great excess of neoblasts above that necessary to ensure survival. In the literature it has been generally assumed that the planarian has

170

a superabundance of these cells, but as a value of $N\alpha = 6$ corresponds to a 99·8 per cent survival probability, this is clearly not the case. In other words, without any accidents, there will still be a 0·2 per cent mortality probability for the animal, due to failure to maintain its own steady state.

The dose-fractionation experiments show that the species *D. etrusca* can repair sub-lethal damage and that one-quarter of the damage (as measured by the LD_{50}) is repaired within the first 12 to 24 hours. As the neoblast cell-cycle time is unlikely to be less than 8 to 12 hours and may be as long as 24 hours, LD_{50} changes during the first 8 to 12 hours are most likely to be a reflection of intracellular repair (Elkind repair). This estimate of neoblast cell-cycle time is based on data from *D. lugubris* on the embryonic neoblast cycle time (Melander 1963), the proportion of neoblast in mitosis, and their rate of increase in the regenerating adult (Lender and Gabriel 1961). The significant increase in LD_{50} between 24 and 48 hours (an interval sufficiently long to allow neoblast pro-liferation) strongly suggests that repopulation is occurring. If this increase is due to repopulation, the process of irradiation can itself result in neoblasts being triggered into cell-cycle. Extrapolation of the exponential rate of decay of residual injury, which is consistent with the data for the first 48 hours after the conditioning dose, indicates that at least 99 per cent of the damage due to the conditioning dose should be repaired within 8 days. After the first day this repair is presumably due to repopulation. However, the low mitotic index found in unirradiated and non-regenerating animals indicates that such turnover rates are not the normal state of affairs.

The significant increase in K (and therefore in D_0) observed even 24 and 48 hours after the initial dose, indicates that neoblasts in the proliferative state (in cell-cycle) are more radioresistant (except for the 8-hour point) than neoblasts at rest. These significant slope changes further support the hypothesis that the first dose decreases the sensitivity (increases the D_0) of the neoblasts to the second radiation dose by increasing the proportion of neoblasts in cell-cycle, some parts of which are more radioresistant.

In recapitulation, the model proposed by Lange and Gilbert (1968) has been further verified by the experimental confirmation of the predicted LD_{50} variation with initial neoblast numbers; it has been shown that the planarian does not possess a superfluity of neoblasts; and the split dose experiments have demon-strated that the neoblasts of *D. etrusca* can repair some sub-lethal damage within hours after irradiation and/or are triggered into cell-cycle (by the initial irradiation), some phases of which are relatively resistant with respect to the resting state.

ADDENDUM

In the tables of papers I and II and succeeding papers of this series, the planarian survival curve parameters are listed under the headings LD_{50} and K.

The survival and mortality curves are defined by the same two parameters (numbers), but

$$K_{survival} = -K_{mortality}$$

with $K_{survival}$ taking negative values only and $K_{mortality}$ taking positive values only.

REFERENCES

Brøndsted, H. V., 1955, *Biol. Rev.*, **30**, 65.

Dubois, F., 1949, *Bull. biol. Fr. Belg.*, **83**, 213.

Elkind, M. M., and Sutton, H., 1959, *Nature, Lond.*, **184**, 1293.

Lange, C. S., 1967, *J. Embryol. exp. Morph.*, **18**, 199; 1968 a, *Expl Gerontol.* **3**, 219; 1968 b, *Int. J. Radiat. Biol.*, **13**, 511; 1968 c, *Ibid.*, **14**, 119; 1968 d, *Ibid.* (in the press).

Lange, C. S., and Gilbert, C. W., 1968, *Int. J. Radiat. Biol.*, **14**, 373.

Lender, T., and Gabriel, A., 1961, *Bull. Soc. zool. Fr.*, **86**, 67.

Melander, Y., 1963, *Hereditas*, **49**, 119.

Verhoef, A. M., 1946, *Proc. K. ned. Akad. Wet.*, **49**, 543.

Studies on the cellular basis of radiation lethality
V. A survival curve for the reproductive integrity of the planarian neoblast and the effect of polyploidy on the radiation response

C. S. LANGE

The experimentally verified model proposed by Lange and Gilbert (1968) quantitatively relates the parameters of the planarian mortality curve ⸱ ⸱ those of the neoblast (planarian stem cell) survival (of reproductive integrity) curve. By use of previously measured values of the constants of the model and the mortality parameters of the members of the *Dugesia lugubris* auto- or allopolyploid series (2*n*, 3*n*, 4*n*), the neoblast survival curves are determined, and the first direct comparison of the effects of polyploidy on radiosensitivity in animal cells is made. No significant effect of polyploidy on radiosensitivity (D_0) was found. The significantly lower survival of triploids with respect to diploids was found to be due to extrapolation number differences. These differences contradict those predicted by target theory. The implications of these findings are also discussed in terms of energy absorption per cell and per chromosome.

1. Introduction

The planarian lends itself to the study of the cellular basis of radiation lethality because it contains a single, morphologically-identifiable population of toti-potential stem cells (cf. Brønsted 1955), without which the animal cannot long survive. The parameters of planarian-mortality probability have been established under conditions where death could only be ascribed plausibly to a cellular mechanism, namely, loss of neoblast (stem cell) reproductive integrity (Lange 1968 b, c). Lange and Gilbert (1968) presented a model which quantitatively relates these parameters (for the organism) to the parameters of the cellular response (neoblast survival of reproductive integrity), and measured the constants and coefficients of this model. Two non-trivial predictions of the model, (1) that planarian-mortality probability would decrease exponentially as the number of reproductively intact neoblasts was increased in an animal of constant volume, and (2) that the radiation LD_{50} would decrease as a logarithmic function of the (decreasing) number of reproductively intact neoblasts in a planarian of constant volume, were tested experimentally and confirmed (Lange and Gilbert 1968, Lange 1968 d). Thus the survival curves of diploid, triploid and tetraploid planarians of the allo- or autopolyploid series of *Dugesia lugubris*, which are derived from the model and the animal-survival data, are of particular interest with respect to the problem of the mechanism of radiation-induced loss of reproductive integrity in animal cells *in vivo*. In this paper, such survival curves are presented, and some of the implications of these findings are discussed.

2. Materials and methods

The materials and methods used to determine the values of the parameters mentioned in this paper can be found, in detail, in one or more of the following references: factors affecting planarian mortality probability (Lange 1968 a, b, c); mortality probability parameters for the ' standard planarian ' (Lange 1968 c); the number of neoblasts in a planarian (Lange 1967, 1968 a); the independent repopulation probability of a single neoblast (Lange and Gilbert 1968); ageing and senescence in the planarian (Lange 1968 a); planarian total tissue volume (Lange 1967, 1968 a); D_0 and the product $N\alpha E$ (Lange and Gilbert 1968).

3. Results—Survival curves for the reproductive integrity of diploid, triploid and tetraploid neoblasts of the planarian *D. lugubris*

The model (Lange and Gilbert 1968) gives the neoblast (stem cell) survival-curve parameters (respectively, the inverse slope and the extrapolated zero dose intercept) D_0 and E in terms of the parameters K and LD_{50} (respectively, the inverse slope and 50 per cent mortality intercept) of the planarian radiation mortality curve as follows:

$$D_0 = \mathrm{K}/1\cdot2 \tag{1}$$

and

$$E = (1/N\alpha)\ (\exp\ (1\cdot2(\mathrm{LD}_{50}/\mathrm{K}) - 0\cdot65)) \tag{2}$$

or

$$E = (1/1\cdot9N\alpha)\ (\exp\ (1\cdot2\mathrm{LD}_{50}/\mathrm{K})), \tag{3}$$

where N is the initial number of neoblasts and α is the independent repopulation probability for a single neoblast.

A primary condition of the model is that the planarian mortality data to be used in the above equations must have been obtained under conditions where the sole probable cause of death was neoblast loss of reproductive integrity. The population standards (a set of planarian mortality curve parameters obtained under specified conditions—also our definition of the ' standard planarian ') given previously (Lange 1968 c), satisfy this condition, and on the basis of those data, values of D_0 and of the product $N\alpha E$ were calculated (Lange and Gilbert 1968).

The number of neoblasts (N_0) in a planarian of median size (10 mm long) was given as $(1\cdot406 \pm 0\cdot267) \times 10^5$ (Lange and Gilbert 1968). This value, based on direct cell counts corrected as described previously (Lange 1967), is valid for diploid and triploid planarians (*D. lugubris*). For the tetraploids, a value of $(1\cdot353 \pm 0\cdot231) \times 10^5$ is obtained from the tetraploid regression line (Lange 1968 a) and Abercrombie's correction factor. Abercrombie (1946) showed that when objects are counted in serial sections, an increase in the ratio of the diameter of the objects to the section thickness is accompanied by an increase in the probability that each object will be counted more than once. Multiplying the crude count by the ratio: object diameter/(object diameter + section thickness), corrects this error. The (nuclear) diameter of the tetraploid neoblasts $(10\cdot9 \pm 0\cdot3\ \mu)$ was found to be significantly larger than that of diploid or triploid neoblasts (respectively, $6\cdot9 \pm 0\cdot2\ \mu$ and $7\cdot0 \pm 0\cdot3\ \mu$), thus resulting in a greater degree of overcounting of the tetraploids. Thus Abercrombie's correction factor was 0·59

174

for diploids and triploids (Lange 1967) but 0·48 for tetraploids. However, the resultant difference in true cell number was found to be not significant.

The value of α, the probability that a single reproductively-intact neoblast can independently repopulate the planarian, was measured in diploid and triploid animals (Lange and Gilbert 1968). As no significant difference was found between the diploid and triploid values, it was assumed that the value for tetraploids does not differ from that of the diploids.

Biotype	$N\alpha E \pm$ S.E.	$N_0 \pm$ S.E.	$\alpha \pm$ S.E.	$N_0\alpha$	$E \pm$ S.E.
Diploid	$20\cdot66 \pm 15\cdot03$	$1\cdot406 \pm 0\cdot267$ $\times 10^5$	$3\cdot76 \pm 0\cdot56$ $\times 10^{-5}$	$5\cdot287$	$3\cdot9 \pm 3\cdot0$
Triploid	$4\cdot48 \pm 2\cdot39$	$1\cdot406 \pm 0\cdot267$ $\times 10^5$	$4\cdot05 \pm 0\cdot83$ $\times 10^{-5}$	$5\cdot694$	$0\cdot79 \pm 0\cdot47$
Tetraploid	$4\cdot61 \pm 3\cdot74$	$1\cdot353 \pm 0\cdot231$ $\times 10^5$	$3\cdot76 \pm 0\cdot56$ $\times 10^{-5}$	$5\cdot087$	$0\cdot91 \pm 0\cdot84$

Table 1. Calculation of the extrapolation number from previously determined parameters (N, α, E, and N_0 are as defined in the text of § 3).

In view of the above findings, the values of the extrapolation numbers (E) can now be calculated from the values of $N\alpha E$, N_0, and α, all of which are given in table 1. The parameters of the survival curves for diploid, triploid and tetraploid neoblasts are given in table 2.

Biotype	$D_0 \pm$ S.E. (rads)	$E \pm$ S.E.
Diploid	727 ± 159	$3\cdot9 \pm 3\cdot0$
Triploid	780 ± 172	$0\cdot79 \pm 0\cdot47$
Tetraploid	983 ± 279	$0\cdot91 \pm 0\cdot84$

Table 2. Parameter values for neoblast survival curves based on independently fitted planarian mortality data.

At this point one may note that when the (mortality) regression lines are independently fitted to the data for each biotype, none of the values of D_0 is significantly different from any other. Since the model relates D_0 directly to K, and if the three biotypes have a common D_0, then it should be possible to fit a set of three parallel lines to the planarian mortality data. This was done. The resulting probit line parameters are given in table 3, and the corresponding neoblast survival curve parameters are presented in table 4. Analysis of variance shows no significant difference between the goodness of fit of the parallel lines to that of the independently fitted lines. Thus the data are consistent with the hypothesis that the degree of polyploidy has no effect on the value of D_0. As a further test of this hypothesis, and in order to see if a dose modification

Biotype	LD$_{50}$ ± S.E. (rads)	K† ± S.E. (rads)	C‡ ± S.E. (per cent)
Diploid	2709 ± 188	961 ± 145	2·7 ± 3·1
Triploid	1674 ± 141	961 ± 145	11·7 ± 5·1
Tetraploid	2246 ± 284	961 ± 145	11·4 ± 10·1

† K = inverse slope.
‡ C = natural mortality.

Table 3. Parameter values for planarian mortality curves fitted with common slope.

mechanism of polyploidy is consistent with the data, lines constrained to have a common zero dose intercept were fitted to the data. This constraint resulted in a significant $(0·05 > P > 0·01)$ increase in variance. Thus polyploidy does not affect neoblast radiosensitivity by a dose modification mechanism; i.e. D_0 is unaffected.

Biotype	NαE ± S.E.	$N_0\alpha$	E ± S.E.	D_0 ± S.E. (rads)
Diploid	15·33 ± 7·63	5·287	2·9 ± 1·6	801 ± 121
Triploid	4·22 ± 1·59	5·694	0·74 ± 0·35	801 ± 121
Tetraploid	8·62 ± 5·30	5·087	1·7 ± 1·1	801 ± 121

Table 4. Calculation of parameter values for neoblast survival curves based on common-slope-fitted planarian mortality data (N, α, E, N_0, and D_0 are as defined in the text of § 3).

The standard errors of the extrapolation number values given in tables 2 and 4 are such that significant differences between them cannot be seen (this is due to the accumulation of errors inherent in an indirect approach). As the value of D_0 does not vary for the members of the auto- or allopolyploid series, then if the extrapolation numbers do not differ significantly, it should be possible to fit a single regression line with common slope and common zero dose intercept to the planarian mortality data. However, the attempt to fit a single common line to the combined data also resulted in a significant increase in variance (see table 5). Thus a common D_0 is consistent with the data, but a common intercept is not! Therefore, the significantly greater LD$_{50}$ of the diploid with respect to the triploid (see table 6) implies that, although no statistically significant difference was found in the calculated extrapolation numbers, the diploid extrapolation number must in fact be greater than that of the triploid. (If this were not the case, a common line for the combined data should fit equally well.) One may justifiably conclude that polyploidy (triploidy and tetraploidy) has no effect on D_0 and that alterations in the surviving fraction of neoblasts are due to differences in extrapolation number (i.e. ' shoulder '). However, the relationship between extrapolation number and ploidy is *not* that predicted by classical target theory.

176

Data set	χ^2	f	Data treatment	χ^2/f	$F_{2,\,72}$	P
Diploid	43·7	18	Population standards	2·4	—	—
Triploid	136·2	49	Data independently	2·8	—	—
Tetraploid	1·1	5	fitted to each line	0·2	—	—
Total	181·0	72	Sum of population standards	2·5	—	—
Pooled data	195·6	74	Pooled data fitted with parallel slope constraint	7·3	2·9	$<0·10$ $>0·05$
Pooled data	202·0	74	Pooled data fitted with common zero-dose intercept constraint	10·5	4·2	$<0·05$ $>0·01$
Pooled data	227·5	76	Pooled data fitted with common line	11·6	4·65	$<0·05$ $>0·01$

Table 5. Analysis of variance showing that a common slope, but *not* a common zero-dose intercept, is consistent with the data for diploid, triploid and tetraploid planarians.

Parameter–parameter (diploid) (triploid)	Curve fitting constraint	t-value	P	Relationship
$LD_{50_{2n}} - LD_{50_{3n}}$	Independent	4·38	$<0·001$	—
	Parallel slope	4·23	$<0·001$	
$P_{1_{2n}} - P_{1_{3n}}$	Independent	1·69	$>0·05$	$P_1 = -K(LD_{50})$
	Parallel slope	4·24	$<0·001$	
$(N\alpha E)_{2n} - (N\alpha E)_{3n}$	Independent	1·06	$>0·05$	$N\alpha E = \exp(1·2P_1 - 0·65)$
	Parallel slope	1·43	$>0·05$	
$E_{2n} - E_{3n}$	Independent	1·04	$>0·05$	$E = N\alpha E/N_0\alpha$
	Parallel slope	1·50	$>0·05$	

Table 6. Significance of diploid–triploid parameter differences.

4. Discussion and conclusions

The model proposed by Lange and Gilbert (1968) defines planarian mortality in terms of the absorbed radiation dose, D, and the four parameters N, α, D_0 and E, the first two of which have been determined independently of the ' standard planarian ' (population standards, see Lange 1968 c) survival experiments, while the latter two have been deduced from the former plus the parameters of the ' standard planarian ' mortality curve. Although this model has been validated by the experimental confirmation of two of its non-trivial predictions (Lange and Gilbert 1968, Lange 1968 d), it should be remembered that these survival curves are based on mortality probability measurements under defined conditions, in which death was, fundamentally, a result of stem-cell sterilization; and the factors which alter the rate of demand for stem-cell differentiation and proliferation were under control. The defined conditions were those of low metabolic rate (and hence low cellular attrition rate) where the planarian neoblasts were given time to repopulate.

No quantitative model yet exists relating the survival time to metabolic rate and/or other physiological parameters. Until it does, it is unlikely that organismal mortality probability will be correctly predicted for all possible metabolic conditions. However, even under the standardized metabolic conditions of the ' standard planarian ', the present model represents the first successful quantitative description of animal mortality probability in terms of cellular reproductive integrity. Unfortunately, direct methods for the study of the neoblast x-ray dose–response curve have so far proved not feasible; should future technical achievements change this situation, we may be able to determine the effects of metabolic and physiological changes on N and α. Such studies would be of great value in the development of a truly quantitative cellular basis for radiation physiology. A recent encouraging development has been the observation of neoblast mitosis *in vitro* reported by Betchaku (1967).

Despite the above limitations, this study has provided a complete determination of the dose–response curve of animal cells under conditions of normal growth regulation in an organized tissue *in vivo*, thereby confirming that the use of the multitarget-type survival curve can be extended to give meaningful results under normal *in vivo* conditions, where cells are not in a state of continuous turnover.

To put these results in perspective, we may examine the other cell systems studied *in vivo*. The first confirmation that survival curves comparable to those found by Puck and Marcus (1956) *in vitro* also obtained *in vivo* was that of Hewitt and Wilson (1959) using a mouse leukaemia. The normal conditions of growth for this cell type, as well as the conditions under which reproductive integrity was assayed, are those of continuous turnover (exponential growth, with a large growth fraction). This system, however, is analogous to the situation *in vitro*, as the host has little or no effect on the growth of the leukaemic cell other than that of a culture chamber filled with nutrient material in an incubator. Hall, Lajtha and Oliver (1962) presented the first survival-curve determination for cells of an organized tissue *in vivo*. But again, in the material studied—the *Vicia* meristem—the normal conditions of growth during the assay period are those of continuous turnover (exponential growth).

The haematopoietic colony-forming cells, studied by Till and McCulloch (1961), have been shown to have survival curves comparable to those of other mammalian cells *in vitro*. These cells are found in mammalian bone-marrow and are presumably part of an organized tissue, but little is known of their normal growth conditions. The assay system in which their reproductive integrity is measured is again one of continuous exponential growth. Also, it is not yet clear what role these cells play in the survival of the animal after irradiation.

Hornsey (1967) has measured the values of $D_f = LD_{50\,fractionated} - LD_{50\,single}$ after dose-fractionation experiments to determine the product $D_0 \ln(E)$ for the stem cells of the gut ($LD_{50/4}$) and bone-marrow ($LD_{50/30}$) in mice. However, until the role of infection in the post-irradiation syndrome can be eliminated, or at least quantitatively defined, the validity of the assumption that

LD$_{50}$ is a directly proportional measure of stem-cell reproductive integrity is very much in doubt. The findings of Byron and Lajtha (1966) indicate that haemato-poietic stem-cell reproductive integrity has little to do with LD$_{50/30}$ if infective commensal organisms are not controlled. The delayed gut death in ' germ-free ' mice (Wilson 1963) may also be a reflection of the absence of infective organisms.

The studies of Withers (1967) do provide direct evidence of the slope of the survival curve for mouse dermis cells, an organized tissue *in vivo*, but owing to the limits of accuracy of the data, and the fact that the data are obtained in regions of heavy depopulation (several decades down the survival curve), extrapolation errors become very large. There is also an appreciable uncertainty as to the size of the cell population at risk. The findings of Kember (1967) for rat growth cartilage are subject to the same limitations as those for mouse dermis, but to an even greater degree. The short range over which survival or reproductive integrity is assayable (only one to two decades) makes the extra-polation even more dubious, and the lack of a sharp end-point makes the data highly variable.

The planarian neoblast is normally a resting cell responding to the needs of the differentiated tissues as they arise. The results of the split-dose experiments with *D. etrusca* (Lange 1968 d) suggest that *some* neoblasts go into cell cycle in response to a dose of radiation. However, the rate of repopulation (decay of residual injury) is such that it is unlikely that all or even most neoblasts are set into cycle. Thus the planarian assay conditions may be very different from those of all the above systems. Therefore, the survival curves determined in this paper are for animal cells of an organized tissue, under conditions of normal growth regulation.

At this point one might note that the value of D_0, 125 ~ 225 rads, for neo-blasts with a rapid turnover (fissiparous species) is similar to those obtained for rapidly growing mammalian cells, whereas the neoblasts with slow turnover rates (*D. lugubris*) have a much higher D_0 (801 rads).

Hall's (1962) method for the determination of the parameters of survival curves of the form:

$$S = 1 - [1 - \exp{(D/D_0)}]^E$$

by matching a single acute dose (D_s) to two equal fractionated doses (D_f) which produce the same biological effect, is of limited use on material in a state of non-continuous turnover, such as the planarian neoblast, for the following reasons. Firstly, a necessary condition for a unique solution is that one of the dose fractions must be small enough to fall on the initial shoulder of the dose–response curve. In the case of the planarian *D. etrusca*, the repair of sub-lethal damage induced by doses of the order likely to fall on the shoulder is unlikely to be detectable so that only the product $D_0 \ln{(E)} = (2D_f - D_s)$ could be determined. Furthermore, this would require not single matched doses, but a whole series of animal survival dose–response curves matched for equal conditioning dose (D_c) and LD$_{50}$/2 (i.e. for $2D_c = $ LD$_{50} = 2D_f$) for comparison with the single acute dose LD$_{50}$ determination. Secondly, a necessary condition for the use of Hall's method is that the fraction interval must be sufficient to allow complete recovery of sub-lethal damage. In his system where cell proliferation is exponential and continuous, the contribution of repopulation to the repair of injury due to D_c is matched by the growth of the control, in contrast to the situation for most organized tissues *in vivo* (non-continuous turnover). Thus systems of the latter type require the solution of the problem, how much of the repair of injury due to D_o is repair of sub-lethal damage and how much is population increase. In the planarian, these two components may be very difficult to separate with great reliability. Thirdly, it is highly likely that in view of such a population's increase between dose fractions in the split -dose group, the two populations to be matched are not in comparable metabolic states. Thus Hall's method is best used on tissues in which the cells are in a state of constant turnover.

Consideration of the values of the parameters of the neoblast survival curves led to the following two comments. Firstly, as a matter of self-consistency, even

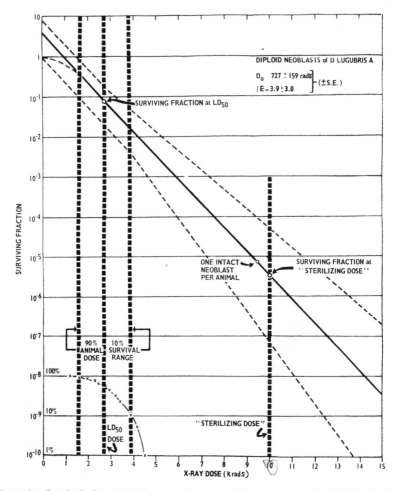

Figure 1. Survival of reproductive integrity curve for the neoblasts of diploid *D. lugubris*. The surviving fraction at 10 krads is less than one intact neoblast per animal, and the 90 per cent to 10 per cent animal survival dose range (the dose range from which this curve is deduced) falls in the first two decades of the survival curve.

Table 7. A summary of the literature of animal polyploidy and radiosensitivity.

Material	Mean chromosome Number	Radiation	n	D_0 (rads)	Reference
Mouse (*Mus musculus*)	Diploid = 40				
Strain L, cell line					
L53	53	^{60}Co γ-rays	2	278±16	
L60	60	,,	2	290±19	
L3	69	,,	2	283±17	Till (1961)
L1	81	,,	2	278±28	
L109	109	,,	2	270±12	
ELD *in vivo*	46	x-rays	(Not a simple function of dose)		Révész and Norman (1960)
ELT *in vivo*	92	,,	,,		
Ehrlich ascites tumour					
Hypodiploid		,,	2·4	109	Silini and Hornsey (1962)
Hypotetraploid		,,	10	114	
Lymphocytic leukaemia					
P-388 'diploid'	40±2·2	3 MeV x-rays	1·6	Oxygenated 160 / Anoxic 360	Berry (1963)
P-288 'tetraploid'		,,	3·1	Oxygenated 180 / Anoxic 380	
Hamster (*Cricetulus griseus*)				(rads)	
♀ lung V79-1 Ovary clone A	22 diploid	55kV x-rays	4·5	128	Elkind (1960)
	22 diploid	,,	6·0–7·1	117	Elkind and Sutton (1960)
Hyperdiploid	23 $2n+1$	^{60}Co γ-rays	2·95	176	Bedford and Hall (1967)
Hypertetraploid	46 $4n+2$	,,	8·50	157	
Artemia salina				(rads)	
Oocytes diploid 1	42	240 kVp x-rays	22·7	422±32	Metalli and Ballardin (1965)
diploid 2	42	,,	69·4	316±82	
tetraploid	84	,,	8·6	1271±248	
Xenopus laevis					
Zygotes haploid	18†	230 kVp x-rays	Curves superimposable if log survival plotted as a function of rads per ploidy		Hamilton (1967)
diploid	36†	,,			
Bombyx mori					
Larvae haploid	29‡	Gaussian sigmoid survival curves like those obtained for whole animals. Stage of development more important than ploidy			Tul'tseva and Astaurov (1958)
diploid	58‡				
triploid	87‡				
Habrobracon juglandis					
Eggs haploid ♂	10§	'No simple correlation between genome number and radiosensitivity throughout the life cycle of *Habrobracon*'			Clark (1957)
Larvae diploid ♀	20§				
Pupae					

† Wickbom (1945).   ‡ Kawaguchi (1928).   § Torvik-Greb (1935).

181

the diploid dose–response curve confirms what was previously assumed—that an x-ray dose of 10 krads sterilizes all the neoblasts in a planarian (i.e. has a reasonable probability of not leaving any reproductively-intact neoblasts in a given animal, see figure 1). Secondly, although the extrapolation number value for the triploids is not significantly different from unity, it might be suggested that the low value may be indicative of a non-reproductively-intact proportion of the cells identified morphologically as neoblasts. Such a hypothesis, however, could not explain an extrapolation number less than unity, since the product $N\alpha$ is independent of the relative size of such a sub-population, and it is this product as the divisor of the term $\exp((LD_{50}/D_0)-0.65)$, which defines the extrapolation number.

The choice of the Planariidae as the experimental material for this study has also made possible a direct test on animal material of the hypothesized importance of the degree of ploidy (i.e. multiplicity of the haploid karyogram) as a factor in determining relative radiosensitivity. An analysis of the literature (see table 7) shows that in most of the experiments performed in attempts to solve this problem, aneuploid or embryonic material has been used. The results of such studies would seem to be irrelevant to this question, because of the effects of polysomy in the first case, and because in the second case, normal development is known to depend on other radiosensitive parameters in addition to the reproductive integrity of the embryo's constituent cells. Thus the only work of direct relevance has been on plants; that of Mortimer (1958) on haploid to hexaploid

Strain	Ploidy	Volume (μ^3)‡	Volume/ploidy‡	E†§	D_0 (krads)†§
Y02022	$n=15$	48·5	48·5	1·0	3·6
Y02587	n	45·2	45·2	1·0	3·6
S163A	n	54·8	54·8	—	—
S216D	n	38·0	38·0	—	—
22D	$2n$ ⎫ homo-zygous	75·5	37·7	2·3 ⎫	9·9 ⎫
87D	$2n$ ⎬	98·5	49·2	1·6 ⎬ avg. 2·3	11·2 ⎬ avg. 11·9
X320	$2n$ ⎰ hetero-zygous	82·5	41·2	3·0 ⎭	14·5 ⎭
X321	$3n$	125	41·6	1·8 ⎫	13·2 ⎫
X322	$3n$	141	47·0	1·8 ⎬ avg. 1·7	14·8 ⎬ avg. 13·6
X354	$3n$	162	54·0	1·7 ⎬	11·2 ⎬
X361	$3n$	154	51·3	1·3 ⎭	15·1 ⎭
X323	$4n$	165	41·2	1·9	10·5
X355	$5n$	225	45·0	1·3	7·2
X362	$6n$	253	42·2	1·4	6·6

† Calculated from photographic enlargements of Mortimer's (1958) published curves.
‡ From Mortimer (1958).
§ 50 kVp x-rays.

Table 8. A comparison of *Saccharomyces cerevisiae* survival-curve parameters† and cell volumes‡ in relation to ploidy.

Saccharomyces, and possibly that of Sparrow and co-workers (Sparrow and Evans 1961, Sparrow and Sparrow 1964, Sparrow, Underbrink and Sparrow 1967) on diploid to dicosaploid *Chrysanthemum* and heptaploid to heptadecaploid *Sedum*. The results of Mortimer for yeast cells, on the one hand, and of Sparrow and co-workers for entire plants, on the other, are in contradiction as to the protective effect of polyploidy. Sparrow *et al.* find a *protective* effect no greater than that to be expected between two different species, one with twice the chromosome number (not *complement*) or interphase chromosome volume of the other (i.e. no difference in energy absorbed per chromosome); whereas Mortimer finds a *sensitizing* effect for ploidy increases above diploid (see table 8 for parameter values calculated from Mortimer's published curves). Although the biological end-point (dose-rate to produce severe growth inhibition) chosen by

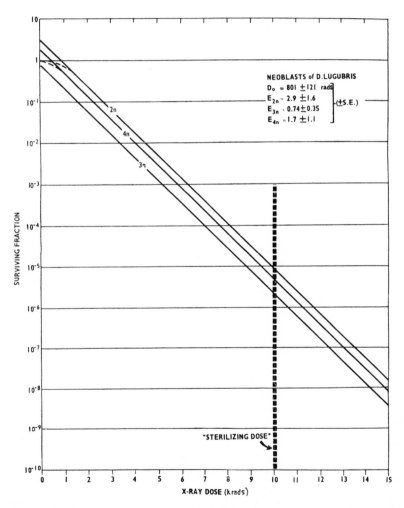

Figure 2.   Survival curves for the neoblasts of diploid, triploid and tetraploid *D. lugubris*.

Sparrow *et al.* is not a direct measure of cellular reproductive integrity, it is related to this parameter, as shown by Hall (1963). Hence the relative sensitivities which they observed may reflect changes in the reproductive integrity of the meristematic cells.

Thus, the hypothesis that redundancy of genetic information protects by providing more 'information replicates' or 'targets' is not supported by the relevant work with plants.

The work of Bedford and Hall (1967) on aneuploid cells of the Chinese hamster (*Cricetulus*) *in vitro* is of interest because their $4n+2$ line contained an exact doubling of the $2n+1$ karyogram of the hyperdiploid line, and no comparable data were available for animal material. Their data support the above protection hypothesis.

For the planarian, *D. lugubris*, the values found for the parameters of the animal survival curves (see tables 2 and 4 and figure 2) show that the triploid is significantly more radiosensitive (lower extrapolation number) than the diploid. Certainly no protective effect of polyploidy was found for the neoblasts. Although the triploid neoblasts were found to have a significantly lower extrapolation number than the diploids (see § 3), the tetraploids were intermediate in shoulder size. The slope of the three survival curves indicates that in terms of

Biotype	Diploid	Triploid	Tetraploid
Chromosome number	8	12	16
D_0 (rads)	801	801	801
E	2·9	0·74	1·7
Neoblast D_{37}[†] (rads)	1620	570	1230
Mean interphase[‡] nuclear volume (μ^3)	1233	1348	5118
Mean interphase[§] chromosome volume (μ^3)	154·2	112·3	319·9
MeV/cell at D_0[‖]	59·2	64·8	246·4
MeV/chromosome at D_0[¶]	7·4	5·4	15·4
Planarian LD_{50} (rads)	2709	1674	2246

[†] From lines of figure 2.

[‡] Calculated from measurements of Lange (1967, 1968 d) corrected for shrinkage and the inclusion of chords in the diameter measurements (MINV).

[§] MINV ÷ chromosome number (MICV).

[‖] MICV × 1·77 ionizations μ^{-3} rads^{-1} × 801 rads × 34 eV (ion pair)$^{-1}$.

[¶] MeV/cell ÷ chromosome number.

Table 9. The discrepancy between energy-absorption changes and cell or animal survival parameters changes with ploidy.

sensitivity per unit dose on the exponential region of the curve, there is no effect of polyploidy. This finding for *D. lugubris* is in general agreement with the findings of Mortimer (1958) for *Saccharomyces cerevisiae* with regard to the absence of a protective effect of polyploidy, but is not in agreement with regard to the effect of polyploidy on slope (D_0). Furthermore, these findings do not agree with those of Sparrow *et al.*, nor of Bedford and Hall (1967), with regard to the protective effect of polyploidy, but are in agreement with regard to the absence of an effect on slope, and in finding the shoulder differences as the reason for differences in relative survival.

One must therefore conclude that at both the level of the organism and at the cellular level, there are factors of much greater importance in the determination of relative radiosensitivity than the degree of polyploidy.

A final point of interest lies in the observation that consideration of the energy per average chromosome deposited after a dose of D_0 (801) rads, places diploid and triploid *D. lugubris* into radiotaxon 8 (7·4 and 5·4 MeV/chromosome, respectively) of the classification of Sparrow *et al.* (1967), whereas the tetraploid is so resistant (15·4 MeV/chromosome) that it might need a new radiotaxon (number 9) to accommodate it. However, the observation that different members of the same auto- or allopolyploid series (*Saccharomyces* and possibly *Dugesia*) fall into different radiotaxa must call into question the meaningfulness of such groupings which are weakly dependent on D_0 and ignore the position parameter (intercept or extrapolation number) of the curve for the survival of reproductive integrity. Table 9 compares the D_0, E, D_{37}, and energy absorption at D_0 of diploid, triploid and tetraploid *D. lugubris* neoblasts. All three have a common D_0, yet energy absorption per cell or per chromosome correlates neither with D_0, nor E, nor D_{37}, nor the LD_{50} of the animal.

ACKNOWLEDGMENTS

I should like to thank Drs. L. G. Lajtha, Alma Howard, C. W. Gilbert, S. Muldal and J. W. Byron for helpful discussions during the course of this work and during the preparation of these papers. I am also indebted to the staff of the Physics Workshop and the Department of Medical Illustration for making and illustrating several pieces of equipment, and to Misses Judith Whelan, Susanne Rutter and Mrs. Pat Nixon for their devoted technical assistance.

REFERENCES

ABERCROMBIE, M., 1946, *Anat. Rec.*, **94**, 239.
BEDFORD, J. S., and HALL, E. J., 1967, *Radiat. Res.*, **31**, 679.
BERRY, R. J., 1963, *Radiat. Res.*, **18**, 236.
BETCHAKU, T., 1967, *J. exp. Zool.*, **164**, 407.
BRØNDSTED, H. V., 1955, *Biol. Rev.*, **30**, 65.
BYRON, J. W., and LAJTHA, L. G., 1966, *Br. J. Radiol.*, **39**, 382.
CLARK, A. M., 1957, *Am. Nat.*, **91**, 111.
ELKIND, M. M., 1960, *Radiology*, **74**, 529.
ELKIND, M. M., and SUTTON, H., 1960, *Radiat. Res.*, **13**, 556.
HALL, E. J., 1962, *Br. J. Radiol.*, **35**, 398; 1963, *Radiat. Res.*, **20**, 195.
HALL, E. J., LAJTHA, L. G., and OLIVER, R., 1962, *Br. J. Radiol.*, **35**, 388.

HAMILTON, L., 1967, *Radiat. Res.*, **30**, 248.
HEWITT, H. B., and WILSON, C. W., 1959, *Br. J. Cancer*, **13**, 69.
HORNSEY, S., 1967, *Radiation Research*, edited by G. Silini (Amsterdam:North Holland Publishing Co.).
KAWAGUCHI, E., 1928, *Zellforsch. m. mikrosk.* **7**, 519.
KEMBER, N. F., 1967, *Br. J. Radiol.*, **40**, 496.
LANGE, C. S., 1967, *J. Embryol. exp. Morph.*, **18**, 199; 1968 a, *Expl. Gerontol.*, **3**, 219; 1968 b, *Int. J. Radiat. Biol.*, **13**, 511; 1968 c, *Ibid.*, **14**, 119; 1968 d, *Ibid.* **14**, 539.

LANGE, C. S., and GILBERT, C. W., 1968, *Int. J. Radiat. Biol.*, **14**, 373.
METALLI, P., and BALLARDIN, E., 1965, Comitato Nazionale Energia Nucleare Report, Roma, November, 1965.
MORTIMER, R. K., 1958, *Radiat. Res.*, **9**, 312.
PUCK, T. T., and MARCUS, P. I., 1956, *J. exp. Med.*, **103**, 653.
RÉVÉSZ, L., and NORMAN, U., 1960, *J. natn. Cancer Inst.*, **25**, 1041.
SILINI, G., and HORNSEY, S., 1962, *Int. J. Radiat. Biol.*, **5**, 147.
SPARROW, A. H., and EVANS, H. J., 1961, *Brookhaven Symp. Biol.*, **14**, 76.
SPARROW, A. H., UNDERBRINK, A. G., and SPARROW, R. C., 1967, *Radiat. Res.*, **32**, 915.
TILL, J. E., 1961, *Radiat. Res.*, **15**, 400.
SPARROW, R. C., and SPARROW, A. H., 1964, *Science, N.Y.*, **147**, 1449.
TILL, J. E., and McCULLOCH, E. A., 1961, *Radiat. Res.*, **14**, 213.
TORVIK-GREB, M., 1935, *Biol. Bull.*, **68**, 25.
TUL'TSEVA, N. M., and ASTAUROV, B. L., 1958, *Biophysics*, **3**, 183.
WICKBOM, T., 1945, *Hereditas*, **31**, 241.
WILSON, B. R., 1963, *Radiat. Res.*, **20**, 477.
WITHERS, H. R., 1967, *Br. J. Radiol.*, **40**, 187.

KEY-WORD TITLE INDEX

AUTHOR INDEX